Math Contests
for
Grades 4, 5, and 6
Volume 4

School Years
1996-1997 through 2000-2001

Written by

Steven R. Conrad • Daniel Flegler

Published by MATH LEAGUE PRESS
Printed in the United States of America

Cover art by Bob DeRosa

Phil Frank Cartoons Copyright © 1993 by CMS

Second Printing, 2007

Copyright © 2001, 2007
by Mathematics Leagues Inc.
All Rights Reserved

Math League Press
P.O. Box 17
Tenafly, NJ 07670-0017

ISBN 0-940805-12-X

Preface

Math Contests—Grades 4, 5, and 6, Volume 4 is the fourth volume in our series of problem books for grades 4, 5, and 6. The first three volumes contain the contests given in the school years 1979-1980 through 1995-1996. Volume 5 contains the contests given in the school years 2001-2002 through 2005-2006. This volume contains contests given from 1996-1997 through 2000-2001. (You can use the order form on page 154 to order any of our 15 books.)

This book is divided into three sections for ease of use by students and teachers. You'll find the contests in the first section. Each contest consists of 30 or 40 multiple-choice questions that you can do in 30 minutes. On each 3-page contest, the questions on the 1st page are generally straightforward, those on the 2nd page are moderate in difficulty, and those on the 3rd page are more difficult. In the second section of the book, you'll find detailed solutions to all the contest questions. In the third and final section of the book are the letter answers to each contest. In this section, you'll also find rating scales you can use to rate your performance.

Many people prefer to consult the answer section rather than the solution section when first reviewing a contest. We believe that reworking a problem when you know the answer (but *not* the solution) often leads to increased understanding of problem-solving techniques.

Each year we sponsor an Annual 4th Grade Mathematics Contest, an Annual 5th Grade Mathematics Contest, and an Annual 6th Grade Mathematics Contest. A student may participate in the contest on grade level or for any higher grade level. For example, students in grades 4 and 5 (or below) may participate in the 6th Grade Contest. Starting with the 1991-92 school year, students have been permitted to use calculators on any of our contests.

Steven R. Conrad & Daniel Flegler, contest authors

Acknowledgments

For demonstrating the meaning of selflessness on a daily basis, special thanks to Grace Flegler.

To Jeannine Kolbush, who did an awesome proofreading job, thanks!

Table Of Contents

The Contests

....

1996-1997 through 2000-2001

4th Grade Contests

1996-1997 through 2000-2001

1996-97 Annual 4th Grade Contest

Spring, 1997

Instructions

4

- **Time** You will have only *30 minutes* working time for this contest. You might be *unable* to finish all 30 questions in the time allowed.

- **Scores** Please remember that *this is a contest, not a test*—and there is no "passing" or "failing" score. Few students score as high as 24 points (80% correct). Students with half that, 12 points, *deserve commendation!*

- **Format and Point Value** This is a multiple-choice contest. Each answer is an A, B, C, or D. Write each answer in the *Answer Column* to the right of each question. A correct answer is worth 1 point. Unanswered questions get no credit. You **may** use a calculator.

1. $(2 + 8) + 10 = 2 \times \underline{?}$

   A) 8   B) 10   C) 12   D) 18

   1. B ✓

2. Of the following, the largest sum is

   A) 1110+9990   B) 1101+9909
   C) 1011+9099   D) 1111+9999

   2. D ✓

3. What number is 10 more than the greatest whole number less than 100?

   A) 99   B) 100   C) 109   D) 110

   3. C ✓

4. The sum $50 + 50 + 50$ is equal to each of the following *except*

   A) 75+75   B) 35+35+35+35
   C) 30+30+30+30+30   D) 25+25+25+25+25+25

   4. B ✓

5. If I have 3 more than 3 dozen eggs, how many eggs do I have?

   A) 24   B) 27   C) 36   D) 39

   5. D ✓

6. $444 + 444 + 444 = (3 \times 400) + (3 \times \underline{?})$

   A) 38   B) 40   C) 42   D) 44

   6. D ✓

7. I had exactly 18 lollipops. If I gave away 4, lost 2, and ate 5, how many lollipops would I have left?

   A) 7   B) 9   C) 11   D) 18

   7. A ✓

8. If today is a Tuesday, 10 days from now is a

   A) Friday   B) Saturday
   C) Sunday   D) Monday

   8. A ✓

9. The largest whole number less than 100 that is a multiple of 8 is

   A) 88   B) 96   C) 98   D) 104

   9. B ✓

10. There are $\underline{?}$ 2-digit whole numbers less than 50.

    A) 50   B) 49   C) 40   D) 39

    10. ⌀C

11. All of the following pairs of numbers have even sums *except*

    A) 676, 989   B) 687, 989   C) 766, 898   D) 898, 988

    11. A ✓

12. A pizza pie is cut into 8 slices. If each slice is cut into 3 pieces, how many pieces of pizza are there altogether?

    A) 11   B) 16   C) 24   D) 38

    12. C ✓

Go on to the next page ⫸ **4**

6

13. If 100 cm = 1 m, then 12 m =

    A) 12 cm   B) 120 cm   C) 1200 cm   D) 12 000 cm

13. ✓
C

14. My 5 quarters are worth as much as ? dimes + 1 nickel.

    A) 6   B) 10   C) 12   D) 13

14. ✓
C

15. Each seat on my school bus holds 2 students. If my bus has 18 seats, then at most ? students can sit on my bus.

    A) 9   B) 16   C) 20   D) 36

15. ✓
D

16. In ? years, I will be 6 years older than I was 2 years ago.

    A) 4   B) 6   C) 8   D) 12

16.
C̸ A

17. Every whole number is divisible by

    A) 3   B) 2   C) 1   D) 0

17.
C ✓

18. The difference between an even and an odd number must be

    A) prime   B) 1   C) even   D) odd

18.
D ✓

19. Double a whole number, then subtract 2. The result is always divisible by

    A) 2   B) 3   C) 4   D) 5

19.
A ✓

20. In which pair of figures below is the number of sides in the first figure one more than the number of sides in the second?

    A) square, triangle   B) square, rectangle
    C) triangle, square   D) rectangle, square

20.
A ✓

21. Ali's 10th birthday will be in 1998. Her 15th birthday will be in

    A) 2002   B) 2003   C) 2012   D) 2013

21. ✓
B

22. This week, you spent $6.25 on lunch and I spent 50¢ less. How much did I spend on lunch?

    A) $1.25   B) $5.75
    C) $5.85   D) $6.75

22.
B ✓

23. I have two coins. If one is worth twice the other, the two coins might be worth ? together.

    A) 6¢   B) 15¢   C) 30¢   D) 35¢

23.
B ✓

Go on to the next page ⫸ **4**

7

24. In HiWay City, odd-numbered routes run North and South, and even-numbered routes run East and West. On how many of the following HiWay City routes can I travel East?

    Route 66 Route 70 Route 89 Route 98

    A) 1   B) 2   C) 3   D) 4

24. C ✓

25. An artist painted 1 happy face yesterday. If he paints twice as many happy faces each day as the day before, how many happy faces will he paint 4 days from now?

    A) 4   B) 8   C) 16   D) 32

25. D ✓

26. If three *different* whole numbers are chosen, one from (6,7,8), one from (2,5,8), and one from (4,6,8), what is the greatest sum that these three numbers could have?

    A) 24   B) 21   C) 20   D) 19

26. B ✓

27. The hundreds' digit of the product $999\,999\,999 \times 888\,888\,888$ is

    A) 1   B) 2   C) 4   D) 9

27. A ✓

28. A square whose side is 12 cm long can be cut up into at most _?_ squares whose sides are each 4 cm long.

    A) 3   B) 6   C) 9   D) 12

28. A C

29. This year on Groundhog Day, the groundhog saw its shadow 120 times. Last year on this day, the groundhog saw its shadow 6 times for every 5 times it saw its shadow this year. How many times did the groundhog see its shadow last year?

    A) 100   B) 144   C) 220   D) 264

29. A B

30. $(2000 + 1999 + \ldots + 1001 + 1000) - (1000 + 999 + \ldots + 2 + 1) = 1000 \times$ _?_

    A) 999   B) 1000
    C) 1001   D) 1002

30. C ✓

The end of the contest 🖎 **4**

Solutions on Page 73 • Answers on Page 138

1997-98 Annual 4th Grade Contest

Spring, 1998

Instructions

- **Time** You will have only *30 minutes* working time for this contest. You might be *unable* to finish all 30 questions in the time allowed.

- **Scores** Please remember that *this is a contest, not a test*—and there is no "passing" or "failing" score. Few students score as high as 24 points (80% correct). Students with half that, 12 points, *deserve commendation!*

- **Format and Point Value** This is a multiple-choice contest. Each answer is an A, B, C, or D. Write each answer in the *Answer Column* to the right of each question. A correct answer is worth 1 point. Unanswered questions get no credit. You **may** use a calculator.

1. What is the sum of 500 ones?

   A) 1   B) 500   C) 501   D) 5000

   1. B ✓

2. 10 hours = ? minutes

   A) 6   B) 60   C) 70   D) 600

   2. D ✓

3. 19 + 98 = 20 + ?

   A) 78   B) 97   C) 99   D) 1978

   3. B ✓

4. When I multiply ? by itself, the result is 256.

   A) 16   B) 32   C) 64   D) 128

   4. A ✓

5. The number two thousand one is a ? -digit number.

   A) 2   B) 3   C) 4   D) 5

   5. C ✓

6. A group of 24 people can be split into ? groups of 3 people each.

   A) 4   B) 6   C) 8   D) 12

   6. C ✓

7. 30 ones has the same value as ? tens.

   A) 3   B) 10   C) 30   D) 300

   7. A ✓

8. (12 ÷ 4) × 3 =

   A) 1   B) 3   C) 6   D) 9

   8. D ✓

9. (10 × 1) + (10 × 10) + (10 × 100) =

   A) 111   B) 1100   C) 1110   D) 1111

   9. C ✓

10. If Gramps has 12 pairs of red socks, he has ? red socks.

    A) 6   B) 12   C) 14   D) 24

    10. D ✓

11. The number of coins in $1.25 worth of nickels is ? more than the number of coins in $1.25 worth of quarters.

    A) 20   B) 21   C) 25   D) 30

    11. A ✓

12. Which of the following is an even number *and* a factor of 18?

    A) 9   B) 6   C) 4   D) 3

    12. B ✓

Go on to the next page ⏭ **4**

10

13. The word *groundbreaking* has _?_ vowels.

   A) 2   B) 3   C) 4   D) 5

13. D ✓

14. $(2 \times 1) + (2 \times 3) + (2 \times 5) =$

   A) 2×8 B) 2×9 C) 6×9 D) $2 + 9$

14. B ✓

15. 7000 cm = _?_ m

   A) 7   B) 70   C) 700   D) 7000

15. B ✓

16. How much greater is 5555 than 1234?

   A) 4021   B) 4123   C) 4231   D) 4321

16. D ✓

17. Since I'm next to last of 8
    people holding a hot dog,
    _?_ people are ahead of me.

   A) 4 B) 5 C) 6 D) 7

17. C ✓

18. How many days in a week have names with more than 6 letters?

   A) 6   B) 5   C) 4   D) 3

18. C ✓

19. If I double _?_ and divide the result by 4, the quotient is 16.

   A) 2   B) 32   C) 64   D) 128

19. B ✓

20. Though I misspelled 5 words on a quiz, I spelled 5 times as
    many correctly. How many words were on the quiz altogether?

   A) 5   B) 10   C) 25   D) 30

20. a D

21. A jigsaw puzzle has 500 pieces,
    of which 85 are edge pieces. How
    many pieces are *not* edge pieces?

   A) 315 B) 415 C) 425 D) 585

21. B ✓

22. $(3 \times 333) + (3 \times 333) = 3 \times$ _?_

   A) 111 B) 333 C) 666 D) 999

22. C ✓

23. When divided by 5, the number _?_ leaves a remainder of 3.

   A) 1998   B) 1999   C) 2001   D) 2002

23. A ✓

Go on to the next page ⟫ **4**

11

24. Tim turned 10 years old exactly two months ago. Tim will turn 12 years old in ? months.

    A) 20   B) 22   C) 24   D) 26

24. B ✓

25. Every number divisible by 8 *must* be divisible by all the following numbers *except*

    A) 6   B) 4   C) 2   D) 1

25. A ✓

26. Divide 891 by 3. The quotient can be divided evenly by

    A) 7   B) 17   C) 27   D) 97

26. C ✓

27. Pat's pot-bellied pig eats 3 pans of pig food a day at a cost of 75¢ per pan. How much does 1 week's worth of pig food cost?

    A) $15.75   B) $5.25   C) $2.25   D) $21

27. A ✓

28. The fewest number of 2's you need to multiply together to get a product greater than 1000 is

    A) 4   B) 9   C) 10   D) 501

28. ⌀ C

29. I read 1 page on March 1, 2 pages on March 2, and so on. Each day in March, I read as many pages as the day number. Altogether, how many pages did I read in March?

    A) 30   B) 31   C) 465   D) 496

29. D ✓

30. To form the number 13 471 897, begin with 1, then 3. Each following digit is the ones' digit of the sum of the two digits before it. A 25-digit number is formed the same way, but starting with 1, then 5. The ones' digit of this number is

    A) 1   B) 5   C) 7   D) 9

30. A C

The end of the contest ✍ **4**

Solutions on Page 77 • Answers on Page 139

12

1998-99 Annual 4th Grade Contest

Spring, 1999

Instructions

4

- **Time** You will have only *30 minutes* working time for this contest. You might be *unable* to finish all 30 questions in the time allowed.

- **Scores** Please remember that *this is a contest, not a test*—and there is no "passing" or "failing" score. Few students score as high as 24 points (80% correct). Students with half that, 12 points, *deserve commendation!*

- **Format and Point Value** This is a multiple-choice contest. Each answer is an A, B, C, or D. Write each answer in the *Answer Column* to the right of each question. A correct answer is worth 1 point. Unanswered questions get no credit. You **may** use a calculator.

Answer Column

1. $1 \times 1 \div 1 \times 1 \div 1 \times 1 \div 1 \times 1 \div 1 \times 1 =$
   A) 0   B) 1   C) 9   D) 10

1. B ✓

2. Anyone eating here needs a fork, a spoon, and a knife. When five of us eat here, the total number of forks plus spoons plus knives we need is
   A) 5   B) 8   C) 10   D) 15

2. D ✓

3. $(1 \times 2 \times 3 \times 4 \times 5) \div 6 =$
   A) 6   B) 20   C) 36   D) 720

3. B ✓

4. Last month, my parrot ate 16 oranges, 12 bananas, and 22 apples. How many pieces of fruit was that?
   A) 40   B) 46   C) 48   D) 50

4. D ✓

5. The sum of the hundreds' digit and the ones' digit of 1999 is
   A) 10   B) 18   C) 28   D) 81

5. A̸ B

6. Each of the following is a whole number *except*
   A) $3 \div 2$ B) $3 - 2$ C) 3×2 D) $3 + 2$

6. A ✓

7. When spelled, how many months have *e* as their second letter?
   A) 1   B) 2   C) 3   D) 4

7. $\frac{1}{3}$ 3 C

8. After I ate four of my two dozen donuts, I had ? donuts left.
   A) 8   B) 16   C) 20   D) 28

8. C ✓

9. $32 \div 4 = 64 \div$?
   A) 2   B) 6   C) 8   D) 16

9. C ✓

10. Michael Jordan's first uniform num-ber was 23. His second was 45. The sum of his two uniform numbers is
    A) 22   B) 68   C) 72   D) 1035

10. B ✓

11. $8 \times 6 \times 4 \times 2 \times 0 =$
    A) 0   B) 1   C) 20   D) 384

11. A ✓

12. What is the correct time exactly 61 minutes after 2:00 P.M.?
    A) 2:59 P.M.   B) 2:61 P.M.   C) 3:01 P.M.   D) 3:31 P.M.

12. C ✓

Go on to the next page ▐▶ **4**

13. The number of consonants in the word *consonant* plus the number of vowels in the word *vowel* equals

A) 3   B) 8   C) 9   D) 11

13. ✓ B

14. My dad found a piggy bank that contained 7 pennies. All the other coins were nickels. The value of the coins in the piggy bank could *not* have been

A) $66.66   B) $67.67
C) $77.77   D) $222.22

14. ✓ A

15. $2+3+4+5+6 = 2\times3\times4\times5\times6 \div \underline{\ ?\ }$

A) 12   B) 24   C) 36   D) 60

15. ✓ C

16. What number is 22 more than the number that is 33 less than 44?

A) 99   B) 55   C) 33   D) 11

16. ✓ C

17. Each of the following products is an even number *except*

A) 11 × 99   B) 44 × 33   C) 55 × 22   D) 88 × 66

17. ✓ A

18. If baseball cards cost fifty cents per pack, how many packs can you buy for five dollars?

A) 5   B) 10   C) 20   D) 50

18. ✓ B

19. The referee is 25 years older than my brother, who is twice my age. If I am 9, how old is the referee?

A) 43   B) 48   C) 50   D) 59

19. ✓ A

20. $7+7+7+7+7+7+7+7 = 7 \times \underline{\ ?\ }$

A) 7   B) 8   C) 49   D) 56

20. ✓ B

21. To compute the *square* of a number, just multiply the number by itself. What is the square of 11?

A) 22   B) 110   C) 111   D) 121

21. ✓ D

22. $1+2+3+4+5+6+7+8 = 11+22+33+44+55+66+77+88 - \underline{\ ?\ }$

A) 10   B) 80   C) 180   D) 360

22. ✓ D

Go on to the next page ⇨ **4**

23. What is the sum of the digits in the number one million?

    A) one   B) one hundred   C) one thousand   D) one million

23. A ✓

24. Alex, born last year, is exactly 4 years younger than Lee. Alex is _?_ days younger than Lee.

    A) 365   B) 1460   C) 1461   D) 1464

24. ✗ C

25. For a class party, we ordered four pizzas in the shape of a square and one pizza in the shape of a pentagon. Added together, the total number of sides that all these pizzas had was

    A) 17   B) 18   C) 20   D) 21

25. D ✓

26. Seven years from now, I will be twice as old as I was one year ago. How old am I now?

    A) 9   B) 8   C) 7   D) 6

26. A ✓

27. If 6 divides evenly into both my age and my grandmother's age, then the sum of our ages could be

    A) 52   B) 54   C) 56   D) 58

27. B ✓

28. Which one is *not* a side of rectangle ABCD?

    A) $\overline{BD}$ B) $\overline{AD}$ C) $\overline{CD}$ D) $\overline{AB}$

28. ✓ A

29. At the canned food drive, my class brought in 5 cans for every 3 cans your class brought in. If your class brought in 60 cans, then my class brought in _?_ cans.

    A) 36   B) 65
    C) 75   D) 100

29. ✗ D

30. Of the whole numbers from 1 through 100, how many are 5 less than another whole number from 1 through 95?

    A) 20   B) 90   C) 95   D) 100

30. ✗ B

The end of the contest 🖎 **4**

Solutions on Page 81 • Answers on Page 140

1999-2000 Annual 4th Grade Contest

Spring, 2000

Instructions

4

- **Time** You will have only *30 minutes* working time for this contest. You might be *unable* to finish all 30 questions in the time allowed.

- **Scores** Please remember that *this is a contest, not a test*—and there is no "passing" or "failing" score. Few students score as high as 24 points (80% correct). Students with half that, 12 points, *deserve commendation!*

- **Format and Point Value** This is a multiple-choice contest. Each answer is an A, B, C, or D. Write each answer in the *Answer Column* to the right of each question. A correct answer is worth 1 point. Unanswered questions get no credit. You **may** use a calculator.

1. What number is 2 more than 999 + 999?

   A) 1001   B) 1998   C) 2000   D) 2002

   1. C ✓

2. I ate 2 more than 2 dozen mints. How many mints did I eat?

   A) 14   B) 22   C) 26   D) 48

   2. C ✓

3. Every day, my elephant begs for 3 bags of peanuts. For how many bags does it beg each week?

   A) 3   B) 7   C) 10   D) 21

   3. D ✓

4. 50 + 100 + 150 = _?_ × 50

   A) 2   B) 3   C) 4   D) 6

   4. D ✓

5. In which of the following numbers is the hundreds' digit greater than the tens' digit?

   A) 9764   B) 8459   C) 1234   D) 1000

   5. A

6. Which of the following sums is *not* equal to the other three?

   A) 19 + 91   B) 18 + 81   C) 27 + 72   D) 36 + 63

   6. A ✓

7. The number _?_ reads the same forwards and backwards.

   A) 98 766 789   B) 45 545 454   C) 12 343 214   D) 10 535 301

   7. A ✓

8. Multiply the number of sides in a triangle by the number of sides in a square. The number _?_ is *not* a factor of this product.

   A) 3   B) 4   C) 5   D) 6

   8. C ✓

9. I ate one frozen yogurt bar for every 3 ice cream bars. If I ate 12 ice cream bars, how many frozen yogurt bars did I eat?

   A) 4   B) 9   C) 15   D) 36

   9. A ✓

10. 21 + 21 + 21 = 31 + 31 + 31 − _?_

    A) 10   B) 11   C) 21   D) 30

    10. D ✓

11. 11+22+33+44 = 1+2+3+4+ _?_

    A) 10   B) 40   C) 100   D) 110

    11. C ✓

12. When 106 is divided by 3, the remainder is

    A) 0   B) 1   C) 2   D) 3

    12. B ✓

Go on to the next page ⟫ **4**

13. Of the following, which is *not* equal to $5 \times 4 \times 3 \times 2$? A) 15×6 B) 8×15 C) 10×12 D) 20×6	13. ✓ A
14. What is the smallest whole number greater than 0 that is divisible by both 8 and 36? A) 4 B) 36 C) 72 D) 288	14. C ✓
15. My school has 5 sets of twins. How many of the students at my school are twins? A) 5 B) 10 C) 15 D) 20	15. ✓ B
16. My initials are the 1st, 12th, and 20th letters of the alphabet, in that order. My name could be A) Alex Louis Thomas B) Amy Lara Sanchez C) Anna Maria Trunk D) Albert Kevin Upton	16. ✓ A
17. Of the 50 whole numbers from 1 to 50, only ? are divisible by 2. A) 23 B) 24 C) 25 D) 26	17. ✓ C
18. 12 tens + 12 ones = 1 hundred + ? ones. A) 12 B) 21 C) 22 D) 32	18. ✓ D
19. The length of one side of a square is 3. The sum of the lengths of the other three sides is A) 3 B) 6 C) 9 D) 12	19. D̸ C
20. Today's *Chef's Specials* were served on 8 red plates, 6 green plates, 5 white plates, and 4 blue plates. How many of today's *Chef's Specials* were *not* served on blue plates? A) 4 B) 18 C) 19 D) 23	20. C ✓
21. The total value of 8 nickels and 7 dimes equals the total value of ? quarters and 7 nickels. A) 2 B) 3 C) 6 D) 8	21. ✓ B
22. ? numbers between 1 and 41 are equal to 5 times an even number. A) two B) four C) eight D) twenty	22. D̸ B

Go on to the next page ⟩⟩⟩ 4

23. A diameter of a circle is twice as long as a side of a square. If a radius of the circle is 2, how long is a side of the square?

A) 1   B) 2   C) 4   D) 8

23.
B

24. When the store security guard retires in the year 2001 A.D., we will be living in the _?_ century.

A) 19th   B) 20th   C) 21st   D) 200th

24.
C ✓

25. When each of the following is divided by 3, the greatest remainder is left by

A) 173   B) 217   C) 364   D) 420

25.
A ✓

26. Add any two odd numbers. The ones' digit of the sum is always

A) 2   B) prime   C) odd   D) even

26.
D ✓

27. Altogether, 12 triangles have as many sides as _?_ rectangles.

A) 4   B) 8   C) 9   D) 16

27.
C ✓

28. My pennies are worth as much as my nickels, my nickels are worth as much as my dimes, and my dimes are worth as much as my quarters. If the value of *all* these coins is $8, how many nickels do I have?

A) 200   B) 40   C) 20   D) 8

28.
B ✓

29. My dog walked 10 times as far as Frank's dog, and Frank's dog walked 10 times as far as Al's dog. If Frank's dog walked 10 km, then my dog walked _?_ km farther than Al's dog.

A) 9   B) 90   C) 99   D) 109

29.
C ✓

30. Four straight lines can cross in as many as six points, as shown. What is the greatest number of points in which five straight lines can cross?

A) 9   B) 10   C) 12   D) 20

30.
AB

The end of the contest 🖎 **4**

Solutions on Page 85 • Answers on Page 141

20

2000-2001 Annual 4th Grade Contest

Spring, 2001

Instructions

4

- **Time** You will have only *30 minutes* working time for this contest. You might be *unable* to finish all 30 questions in the time allowed.

- **Scores** Please remember that *this is a contest, not a test*—and there is no "passing" or "failing" score. Few students score as high as 24 points (80% correct). Students with half that, 12 points, *deserve commendation!*

- **Format and Point Value** This is a multiple-choice contest. Each answer is an A, B, C, or D. Write each answer in the *Answer Column* to the right of each question. A correct answer is worth 1 point. Unanswered questions get no credit. You **may** use a calculator.

	Answer Column
1. $20 \times 10 \times 2 \times 1 =$ A) 212 B) 221 C) 400 D) 420	1.
2. What is 1 more than 20 more than 300? A) 32 B) 51 C) 312 D) 321	2.
3. $9 + 1 + 8 + 2 + 7 + 3 = 6 + \underline{?}$ A) 4 B) 24 C) 30 D) 36	3.
4. Ten quarters is worth as much as $\underline{?}$ dimes. A) 5 B) 10 C) 15 D) 25	4.
5. A period of $\underline{?}$ weeks is exactly 56 days long. A) 6 B) 7 C) 8 D) 9	5.
6. How many of the letters in the word *mathematics* are vowels? A) 11 B) 7 C) 4 D) 3	6.
7. Which number is twenty thousand, one hundred one? A) 2101 B) 20 101 C) 21 101 D) 201 001	7.
8. 3 hours = 2 hours + 10 minutes + $\underline{?}$ minutes A) 30 B) 50 C) 60 D) 90	8.
9. The number that's 10 less than 2001 is 10 more than A) 1981 B) 1991 C) 2001 D) 2011	9.
10. At the picnic, Sue swallowed 1 of every 6 seeds in her slice of watermelon. Sue must have swallowed $\underline{?}$ of the 162 seeds in her slice. A) 27 B) 28 C) 52 D) 156	10.
11. $101 \times 10 \times 1 \times 0 \times 1 \times 10 \times 101 = 1010 \times \underline{?}$ A) 0 B) 1 C) 2 D) 3	11.
12. $21 \times 21 = 7 \times 7 \times \underline{?}$ A) 3 B) 7 C) 9 D) 21	12.

PICNIC

Go on to the next page ⫸ **4**

22

2000-2001 4TH GRADE CONTEST

13. A triple is a group of 3, so 1 dozen pairs equals __?__ triples. A) 6 B) 8 C) 12 D) 36	13.
14. The sum of the lengths of all the sides of a square equals the length of 1 side multiplied by A) 2 B) 3 C) 4 D) 8	14.
15. It took FedStork 11 weeks + 11 days to deliver the medical bag. How many hours was that? A) 22 B) 88 C) 121 D) 2112	15.
16. If 2 rocks are worth 5 stones, then __?__ stones are worth 10 rocks. A) 2 B) 4 C) 25 D) 50	16.
17. By how many 8s does 32+32+32+32 exceed 24+24+24+24? A) 4 B) 8 C) 24 D) 32	17.
18. (# sides in a rectangle) − (# sides in a square) = (# of sides in a triangle) − __?__ A) 0 B) 1 C) 2 D) 3	18.
19. 145 + 146 + 147 + 148 + 149 = 5× __?__ A) 146 B) 147 C) 148 D) 149	19.
20. Of the 101 times I squirted Dad with a hose last summer, the number of times I missed is 31 more than the number of times I got him wet. How many of the 101 squirts missed Dad? A) 35 B) 49 C) 66 D) 70	20.
21. Of the following sums, which is *not* divisible by 3 + 3 + 3? A) 33 + 33 B) 6 + 6 + 6 + 6 + 6 + 6 C) 9 + 9 + 9 + 9 + 9 D) 18 + 18	21.
22. What is the sum of the remainders of 1234 ÷ 5 and 6789 ÷ 10? A) 4 B) 5 C) 9 D) 13	22.

Go on to the next page ⎯➤ **4**

23

23. The product of a whole number and itself must be divisible by

    A) 1, but not necessarily 2   B) 2, but not necessarily 1
    C) 1 & 2, but not necessarily 3   D) 1, 2, & 3

    23.

24. (sum of all digits in 2000) × (sum of all digits in 2001) =

    A) 0   B) 2   C) 3   D) 6

    24.

25. A dinosaur grows 2 m each week. A
    dragon grows 1 m each day. In 4
    weeks, how much more does a
    dragon grow than a dinosaur?

    A) 4   B) 8   C) 12   D) 20

    25.

26. The sum of the lengths of 4 diameters of a
    circle is 128. How long is a radius of this circle?

    A) 4   B) 8   C) 16   D) 32

    26.

27. If they may touch but not overlap, at most how many squares
    of area 4 can fit inside a rectangle with width 6 and length 14?

    A) 14   B) 21   C) 28   D) 49

    27.

28. An *acronym* is a word formed from the first one or more letters
    of each word in a group of words. If "UFO" is an acronym for
    "unidentified flying object," then for how many of the
    following word groups could MATH be an acronym?
      I.   Multiply All Those Hundreds
      II. MArtians Take Hostages
      III. MAThew Hides
      IV. Minutes After The Hour

    A) one   B) two   C) three   D) four

    28.

29. Each of the 5 target stripes on my shirt is worth
    a different odd whole number less than 50. What
    is the greatest possible sum of these 5 numbers?

    A) 220   B) 225   C) 235   D) 245

    29.

30. 2001 + (2000 − 1999 + 1998 − 1997 + 1996 − . . . + 2 − 1) =

    A) 2001   B) 3001   C) 4001   D) 4002

    30.

The end of the contest ✍ **4**

Solutions on Page 89 • Answers on Page 142

5th Grade Contests

1996-1997 through 2000-2001

1996-97 Annual 5th Grade Contest

Spring, 1997

Instructions

5

- **Time** Do *not* open this booklet until you are told by your teacher to begin. You will have only *30 minutes* working time for this contest. You might be *unable* to finish all 30 questions in the time allowed.

- **Scores** Please remember that *this is a contest, not a test*—and there is no "passing" or "failing" score. Few students score as high as 24 points (80% correct). Students with half that, 12 points, *should be commended!*

- **Format and Point Value** This is a multiple-choice contest. Each answer is an A, B, C, or D. Write each answer in the *Answer Column* to the right of each question. A correct answer is worth 1 point. Unanswered questions get no credit. You **may** use a calculator.

1. $5 + 10 + 15 = 4 + 9 + 14 +$?

   A) 2   B) 3   C) 4   D) 5

   1.

2. I looked in a hole to find your lost marbles and I found 3 blue, 4 red, 2 white, and 5 green marbles. How many more of these marbles were *not* green than were green?

   A) 4   B) 5   C) 9   D) 14

   2.

3. 5 nickels + 5 dimes has the same value as ? quarters.

   A) 2   B) 3   C) 4   D) 5

   3.

4. $111 - 11 - 1 = (777 - 77 - 7) \div$?

   A) 7   B) 8   C) 9   D) 10

   4.

5. The product of two even numbers and one odd number is

   A) even   B) odd   C) more than 10 D) prime

   5.

6. 19 hundreds + 8 tens + 17 ones =

   A) 1987   B) 1996   C) 1997   D) 2717

   6.

7. If my art class starts at 2:45 P.M. and ends at 4:15 P.M., then my art class is ? minutes long.

   A) 60   B) 75   C) 90   D) 120

   7.

8. $500 \text{ cm} + 50 \text{ m} + 5 \text{ km} =$? m

   A) 55   B) 550   C) 555   D) 5055

   8.

9. If the number of marchers in the last parade was equal to the largest even number less than 2000, how many people marched in that parade?

   A) 1000   B) 1998   C) 1999   D) 2000

   9.

10. How many hours are in one week?

    A) 7   B) 24   C) 168   D) 10 080

    10.

11. A Pizza Heaven pizza costs $10.80 and is always cut into 8 equal slices. What is the cost per slice of Pizza Heaven pizza?

    A) $1.10   B) $1.20   C) $1.25   D) $1.35

    11.

Go on to the next page IIII➤ **5**

28

12. $6 \times (3 \times 5 \times 7) - (3 \times 5 \times 7) = \underline{?} \times (3 \times 5 \times 7)$

    A) 7   B) 6   C) 5   D) 4

12.

13. What is the ones' digit of the product $2 \times 2 \times 2 \times 2 \times 5 \times 5 \times 5 \times 5$?

    A) 0   B) 1   C) 2   D) 5

13.

14. The value of 1 quarter + 9 dimes + 9 nickels + 7 pennies is

    A) $1.67   B) $1.97   C) $2.52   D) $19.97

14.

15. In total, how many 3-digit whole numbers use *all three* of the digits 1, 2, and 3?

    A) 3   B) 4   C) 5   D) 6

15.

16. I tried to hit a tennis ball moving at 240 km/hr. That equals a speed of $\underline{?}$ km/min.

    A) 2   B) 4   C) 6   D) 8

16.

17. Of the following, the largest product is

    A) $2 \times 4 \times 9$ B) $2 \times 3 \times 12$
    C) $3 \times 4 \times 6$ D) $3 \times 5 \times 5$

17.

18. The perimeter of one square is twice that of another. If a side of the larger square is 6 cm long, then the area of the smaller is

    A) 9 cm^2 B) 18 cm^2 C) 36 cm^2 D) 72 cm^2

18.

19. When filled, a tray holds $10 worth of quarters. How many quarters are needed to fill this tray?

    A) 25   B) 40   C) 50   D) 250

19.

20. My aunt wears a great many hats. If the number of hats my aunt wears is 1 less than the thousands' digit of 17 854, then my aunt wears $\underline{?}$ hats.

    A) 9   B) 8   C) 7   D) 6

20.

21. If Jo is the 2nd tallest of 126 fifth grade students, how many of these students are shorter than Jo?

    A) 123   B) 124   C) 125   D) 126

21.

22. How much must I pay for a 5-minute phone call that costs $1 for the first minute and 75¢ for each additional minute?

    A) $3.25   B) $3.75   C) $4   D) $5

22.

Go on to the next page ▐▶ **5**

29

23. In which of the following divisions is the remainder 1 more than the remainder you get when you divide 176 by 3?

    A) 173 ÷ 5   B) 174 ÷ 4   C) 175 ÷ 3   D) 176 ÷ 2

    23.

24. I'm thinking of 3 unequal whole numbers, each less than 10. Of the following, each could be their sum *except*

    A) 4   B) 6   C) 24   D) 25

    24.

25. At a book sale, if I sold half the books, you sold half the remainder, and 40 books were not sold, then we started with ? books.

    A) 10   B) 80   C) 120   D) 160

    25.

26. Last week, I bought 1 pog (my first pog). The number I buy doubles each week. If I keep all my pogs, 3 weeks from now I'll have a total of ? pogs.

    A) 31   B) 16   C) 15   D) 7

    26.

27. If a rectangular window which is twice as high as it is wide has a width of 30 cm, then the sum of its width and height is

    A) 45 cm   B) 60 cm   C) 90 cm   D) 120 cm

    27.

28. If 2 of every 3 scarecrows dance the macarena, how many of a group of 120 scarecrows *don't* do this dance?

    A) 30   B) 40   C) 60   D) 80

    28.

29. In counting from 1 to 1000, at some point I call out 5 numbers in a row whose sum is 600. The smallest of these 5 numbers is

    A) 596   B) 120
    C) 118   D) 116

    29.

30. Jack added all the odd numbers from 1 to 1999. Jill added all the even numbers from 2 to 2000. Jill's sum is ? more than Jack's.

    A) 999   B) 1000   C) 1999   D) 2000

    30.

The end of the contest ✍ **5**

Solutions on Page 95 • Answers on Page 143

1997-98 Annual 5th Grade Contest

Spring, 1998

Instructions

5

- **Time** Do *not* open this booklet until you are told by your teacher to begin. You will have only *30 minutes* working time for this contest. You might be *unable* to finish all 30 questions in the time allowed.

- **Scores** Please remember that *this is a contest, not a test*—and there is no "passing" or "failing" score. Few students score as high as 24 points (80% correct). Students with half that, 12 points, *should be commended!*

- **Format and Point Value** This is a multiple-choice contest. Each answer is an A, B, C, or D. Write each answer in the *Answer Column* to the right of each question. A correct answer is worth 1 point. Unanswered questions get no credit. You **may** use a calculator.

Copyright © 1998 by Mathematics Leagues Inc.

1. $5 + 10 + 15 + 20 + 25 = (1 + 2 + 3 + 4 + 5) \times \underline{\ ?\ }$

    A) 2   B) 3   C) 4   D) 5

    1.

2. A certain basket can hold a half-dozen eggs. How many of these baskets do I need to hold 3 dozen eggs?

    A) 3   B) 4   C) 5   D) 6

    2.

3. What is the measure of the largest angle in a right triangle?

    A) 45°   B) 60°   C) 90°   D) 180°

    3.

4. Of the following, which has a value different from the others?

    A) 40×50 B) 4×5000 C) 50×400 D) 40×500

    4.

5. What is the remainder when 222 222 222 is divided by 4?

    A) 3   B) 2   C) 1   D) 0

    5.

6. If one bag of chips costs 75¢, then three of these bags cost

    A) $0.25   B) $1.50   C) $2.25   D) $3.00

    6.

7. Of the following, which has the largest odd factor?

    A) 30   B) 32   C) 36   D) 40

    7.

8. My baby brother will be 3 weeks old in 2 days. How many days old is he today?

    A) 18   B) 19   C) 20   D) 21

    8.

9. $10 + 11 + 12 + 13 + 14 =$
   $(30 + 31 + 32 + 33 + 34) - \underline{\ ?\ }$

    A) 5   B) 20   C) 3×20 D) 5×20

    9.

10. 1 thousand + 9 hundreds + 8 tens + 18 ones =

    A) 1918   B) 1988   C) 1998   D) 19 818

    10.

11. Which of the following quotients is 1 more than $162 \div 18$?

    A) $128 \div 16$ B) $120 \div 15$ C) $132 \div 12$ D) $110 \div 11$

    11.

12. (number of angles in a square) + (number of angles in a triangle) =

    A) 7   B) 9   C) 10   D) 12

    12.

Go on to the next page ⫸ **5**

13. The product of any 3-digit whole number and any 2-digit whole number can contain at most how many digits?

    A) 3   B) 4   C) 5   D) 6

14. The largest whole-number multiple of 7 less than 200 is

    A) 187   B) 189   C) 196   D) 197

15. Dancing pencils cost 74¢ each for the first dozen and 69¢ each for the rest. How much will it cost to buy a dancing pencil for each of your 27 friends?

    A) $18.63 B) $19.23 C) $19.38 D) $19.98

16. (The number of seconds in a week) ÷ (the number of minutes in a week) =

    A) 60   B) 420   C) 3600   D) 7200

17. If 60 cm of snow falls each hour, how much falls in 100 minutes?

    A) 90 cm   B) 1 m   C) 110 cm   D) 120 cm

18. If the sum of 9 numbers is 1998, then their average is

    A) 9 + 1998   B) 9 × 1998   C) 1998 ÷ 9   D) 9 ÷ 1998

19. How many times does the second hand go completely around the face of her circular clock while Mom jogs each morning for 5 minutes?

    A) 5   B) 30   C) 60   D) 300

20. A polygon that contains exactly 5 angles has exactly ? sides.

    A) 3   B) 5   C) 8   D) 10

21. What is the product of the least common multiple of 6 and 18 and the greatest common factor of 6 and 18?

    A) 6   B) 18   C) 54   D) 108

22. A rectangle 2 cm wide and 4 cm long can be divided into how many squares with sides 1 cm long?

    A) 2   B) 4   C) 6   D) 8

Go on to the next page ⏩ **5**

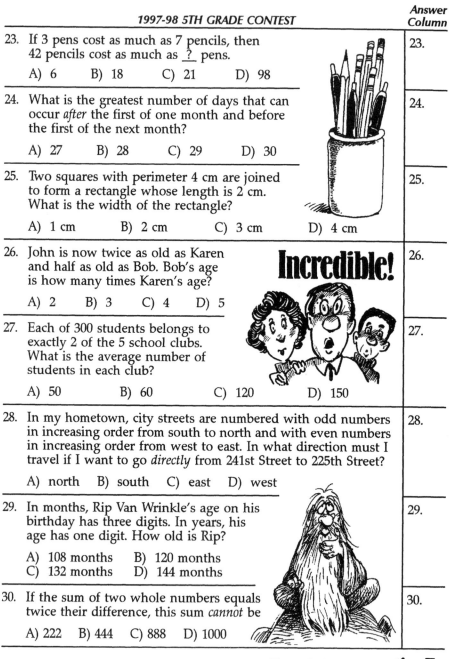

23. If 3 pens cost as much as 7 pencils, then 42 pencils cost as much as _?_ pens.

    A) 6   B) 18   C) 21   D) 98

    23.

24. What is the greatest number of days that can occur *after* the first of one month and before the first of the next month?

    A) 27   B) 28   C) 29   D) 30

    24.

25. Two squares with perimeter 4 cm are joined to form a rectangle whose length is 2 cm. What is the width of the rectangle?

    A) 1 cm   B) 2 cm   C) 3 cm   D) 4 cm

    25.

26. John is now twice as old as Karen and half as old as Bob. Bob's age is how many times Karen's age?

    A) 2   B) 3   C) 4   D) 5

    26.

27. Each of 300 students belongs to exactly 2 of the 5 school clubs. What is the average number of students in each club?

    A) 50   B) 60   C) 120   D) 150

    27.

28. In my hometown, city streets are numbered with odd numbers in increasing order from south to north and with even numbers in increasing order from west to east. In what direction must I travel if I want to go *directly* from 241st Street to 225th Street?

    A) north   B) south   C) east   D) west

    28.

29. In months, Rip Van Wrinkle's age on his birthday has three digits. In years, his age has one digit. How old is Rip?

    A) 108 months   B) 120 months
    C) 132 months   D) 144 months

    29.

30. If the sum of two whole numbers equals twice their difference, this sum *cannot* be

    A) 222   B) 444   C) 888   D) 1000

    30.

The end of the contest 5

1998-99 Annual 5th Grade Contest

Spring, 1999

Instructions

5

- **Time** Do *not* open this booklet until you are told by your teacher to begin. You will have only *30 minutes* working time for this contest. You might be *unable* to finish all 30 questions in the time allowed.

- **Scores** Please remember that *this is a contest, not a test*—and there is no "passing" or "failing" score. Few students score as high as 24 points (80% correct). Students with half that, 12 points, *should be commended!*

- **Format and Point Value** This is a multiple-choice contest. Each answer is an A, B, C, or D. Write each answer in the *Answer Column* to the right of each question. A correct answer is worth 1 point. Unanswered questions get no credit. You **may** use a calculator.

1. 1999 + 2001 + 1999 + 2001 + 1999 + 2001 + 1999 + 2001 =

    A) 4000   B) 8000   C) 12000   D) 16000

    1.

2. One ten plus two hundreds plus three ones equals

    A) 123   B) 213   C) 231   D) 312

    2.

3. Of the following, which is a whole number?

    A) 25 ÷ 3   B) 26 ÷ 3   C) 27 ÷ 3   D) 28 ÷ 3

    3.

4. At *Soup From Doc*, it costs me a quarter for each tablespoon of soup I buy. How many more quarters do I need to buy seven tablespoons of soup than I need to buy four?

    A) 3   B) 7   C) 75   D) 175

    4.

5. What is the quotient of the division $(2 \times 4 \times 6 \times 8) \div (1 \times 2 \times 3 \times 4)$?

    A) 2   B) 4   C) 8   D) 16

    5.

6. The ones' digit of $56 \times 67 \times 78$ is

    A) 8   B) 6   C) 4   D) 1

    6.

7. If yesterday were Monday, then 8 days from today would be

    A) Monday   B) Tuesday   C) Wednesday D) Thursday

    7.

8. The ten-thousands' digit of 654321 + the tens' digit of 654321 =

    A) 8   B) 7   C) 6   D) 5

    8.

9. A year's supply of Frisbees cost me $48.72. My average monthly cost for Frisbees is

    A) $4.06   B) $4.60   C) $4.66   D) $6.00

    9.

10. What is the product of the quotient and the remainder when 1111 is divided by 22?

    A) 25   B) 250   C) 550   D) 1100

    10.

11. Add the total number of sides in one tri-angle, one rectangle, and one hexagon.

    A) 3   B) 11   C) 12   D) 13

    11.

Go on to the next page ‖➡ **5**

36

12. When 10 000 is divided by 9, the remainder is

    A) 1   B) 3   C) 5   D) 7

12.

13. Polly was born on the 200th day of the year. Her birthday falls in

    A) June   B) July
    C) August   D) September

13.

14. 753 is 357 more than

    A) 396 B) 404 C) 406 D) 1110

14.

15. 13 hundreds + 13 tens + 13 ones =

    A) 333 B) 1333 C) 1433 D) 1443

15.

16. The average of two odd numbers is always

    A) odd   B) even   C) prime   D) whole

16.

17. Of the following, which is worth the most?

    A) 45 nickels B) 11 dimes C) 5 quarters D) 1 dollar

17.

18. If I spend one-third of my $120 gift, I'll have _?_ left.

    A) $40   B) $60   C) $80   D) $90

18.

19. The difference between Grandpa's height and mine is 123 cm. If Grandpa's height is 202 cm, then my height is _?_ cm.

    A) 77 B) 79 C) 87 D) 89

19.

20. I took a 7-week calendar and colored in every day that began with an S or a T. How many days stayed uncolored?

    A) 49 B) 35 C) 28 D) 21

20.

21. The first of two numbers is 17 more than twice the second. If the first is 23, what is the sum of the two numbers?

    A) 26   B) 40   C) 73   D) 96

21.

22. Of the following, which has more different whole number factors than the other three?

    A) 4   B) 6   C) 9   D) 25

22.

Go on to the next page ⟶ **5**

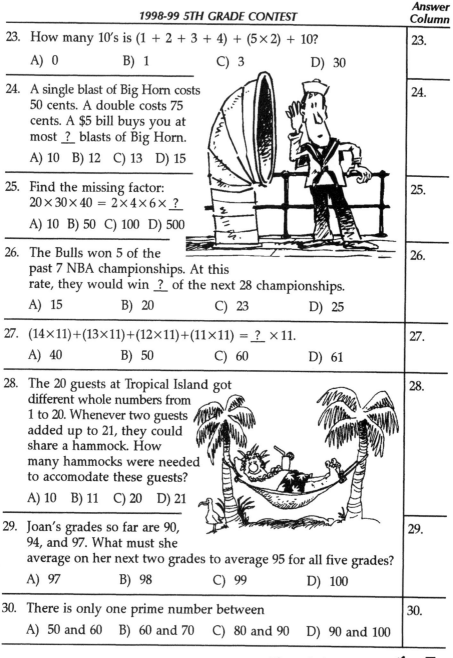

23. How many 10's is $(1 + 2 + 3 + 4) + (5 \times 2) + 10$?

   A) 0   B) 1   C) 3   D) 30

23.

24. A single blast of Big Horn costs 50 cents. A double costs 75 cents. A $5 bill buys you at most _?_ blasts of Big Horn.

   A) 10   B) 12   C) 13   D) 15

24.

25. Find the missing factor:
   $20 \times 30 \times 40 = 2 \times 4 \times 6 \times \underline{?}$

   A) 10   B) 50   C) 100   D) 500

25.

26. The Bulls won 5 of the past 7 NBA championships. At this rate, they would win _?_ of the next 28 championships.

   A) 15   B) 20   C) 23   D) 25

26.

27. $(14 \times 11) + (13 \times 11) + (12 \times 11) + (11 \times 11) = \underline{?} \times 11$.

   A) 40   B) 50   C) 60   D) 61

27.

28. The 20 guests at Tropical Island got different whole numbers from 1 to 20. Whenever two guests added up to 21, they could share a hammock. How many hammocks were needed to accomodate these guests?

   A) 10   B) 11   C) 20   D) 21

28.

29. Joan's grades so far are 90, 94, and 97. What must she average on her next two grades to average 95 for all five grades?

   A) 97   B) 98   C) 99   D) 100

29.

30. There is only one prime number between

   A) 50 and 60   B) 60 and 70   C) 80 and 90   D) 90 and 100

30.

The end of the contest ✍ **5**

1999-2000 Annual 5th Grade Contest

Spring, 2000

5

Instructions

- **Time** Do *not* open this booklet until you are told by your teacher to begin. You will have only *30 minutes* working time for this contest. You might be *unable* to finish all 30 questions in the time allowed.

- **Scores** Please remember that *this is a contest, not a test*—and there is no "passing" or "failing" score. Few students score as high as 24 points (80% correct). Students with half that, 12 points, *should be commended!*

- **Format and Point Value** This is a multiple-choice contest. Each answer is an A, B, C, or D. Write each answer in the *Answer Column* to the right of each question. A correct answer is worth 1 point. Unanswered questions get no credit. You **may** use a calculator.

Answer Column

1. $12 + 14 + 16 + 18 = 2 + 4 + 6 + 8 + \underline{\ ?\ }$
   A) 10   B) 20   C) 30   D) 40

   1.

2. I'm wearing a chain and pendant that I got 17 days before Tuesday. What day was that?
   A) Thur.   B) Fri.   C) Sat.   D) Sun.

   2.

3. $100 \div 5 = \underline{\ ?\ } \times 5$
   A) 2   B) 4   C) 10   D) 20

   3.

4. What is the product of the 5 smallest whole numbers?
   A) 0   B) 15   C) 120   D) 121

   4.

5. The ones' digit of 246 810 is $\underline{\ ?\ }$ less than its hundreds' digit.
   A) 0   B) 1   C) 7   D) 8

   5.

6. Two whole numbers whose difference is odd must have $\underline{\ ?\ }$ sum.
   A) a one-digit B) a prime   C) an even   D) an odd

   6.

7. The value of 5 quarters is the same as the value of $\underline{\ ?\ }$ nickels.
   A) 15   B) 20   C) 25   D) 50

   7.

8. Gil the Fish weighs twice as much as Bill the Fisherman. If Gil weighs 150 kg, then Bill weighs $\underline{\ ?\ }$ kg.
   A) 75 B) 150 C) 225 D) 300

   8.

9. $2 + 22 + 222 = 2 \times \underline{\ ?\ }$
   A) 1 + 11 + 111   B) 1 + 10 + 110
   C) 2 + 12 + 112   D) 1 + 12 + 24

   9.

10. (number of digits in ten million) ÷ (number of digits in one thousand) =
    A) 2   B) 8/3   C) 4   D) 10 000

    10.

11. Add the number of sides in a triangle to the number of sides in a pentagon. The sum equals the number of sides in
    A) a square   B) a rhombus C) a hexagon   D) an octagon

    11.

12. If my secret number uses the digits 1, 2, and 3 once each, in some order, then you can guess my number in at most $\underline{\ ?\ }$ tries.
    A) 3   B) 4   C) 5   D) 6

    12.

Go on to the next page ⟫ **5**

40

13. Which number exceeds the difference between 300 and 100 by 500?

   A) 200   B) 300   C) 600   D) 700

14. 2 hours − ? minutes = 45 minutes

   A) 30   B) 75   C) 85   D) 155

15. Kyle cried 3 crocodile tears each day. How many tears did Kyle cry last week?

   A) 7   B) 10   C) 15   D) 21

16. Add 555 555 555 555 555 to itself. How many 0s appear in the sum?

   A) 1   B) 5   C) 15   D) 29

17. The ones' digit of the largest multiple of 7 that's less than 1000 is

   A) 3   B) 4   C) 7   D) 9

18. When ? is divided by 6, the remainder is 1.

   A) 612 481 230   B) 612 481 239   C) 612 481 238   D) 612 481 237

19. The average of 2000 fours equals the average of 1000 ? .

   A) twos   B) fours   C) sevens   D) eights

20. How much longer is a side of a square with perimeter 36 cm than the width of a rectangle with area 36 cm^2 and length 18 cm?

   A) 4 cm   B) 7 cm   C) 32 cm   D) 34 cm

21. How many kids are in the chorus if no one is younger than 10 years old, 20 kids are 10 or younger, 8 are older than 10, and 6 are older than 11?

   A) 22   B) 26   C) 28   D) 34

22. The number ? has more than two whole-number factors.

   A) 11   B) 13   C) 15   D) 17

23. $40-39+38-37+36-35+34-33+\ldots-9+8-7+6-5+4-3+2-1 =$

   A) 1   B) 20   C) 21   D) 40

Go on to the next page ⮕ **5**

24. In square *ABCD*, if *AB* = 5, then *AB* + *BC* =

    A) *AD* + *AC*   B) *BC* + *BD*
    C) *AC* + *BC*   D) *AD* + *AB*

    24.

25. The lengths of three sides of a ? could be 8 cm, 8 cm, and 16 cm.

    A) rectangle   B) square   C) triangle   D) circle

    25.

26. A different whole number from 1 to 26 is assigned to each letter of the alphabet and written on the kindergarten's 26 alphabet blocks. What is the sum of the numbers assigned to the consonants plus the sum of the numbers assigned to the vowels?

    A) 300   B) 326   C) 330   D) 351

    26.

27. If the value of my seven coins is 57¢, I have exactly one

    A) penny   B) nickel   C) dime   D) quarter

    27.

28. The difference between 19 992 000 and some smaller whole number equals the difference between some larger whole number and 19 992 000. The average of the smaller and larger whole numbers is

    A) 6 664 000   B) 9 996 000   C) 19 992 000   D) 39 984 000

    28.

29. Seven identical 2 m tall ice sculptures were carved so that one was completed at the end of each hour in a 7-hour period of time. Each sculpture shrank at the rate of 10 cm per hour. When completed, the last sculpture was ? cm taller than the first sculpture was at that very moment.

    A) 50   B) 60   C) 70   D) 80

    29.

30. Five boys and four girls are standing in a circle. Just two of the boys can say "Next to me is a boy." How many of the girls can say "Next to me is a girl"?

    A) 0   B) 1   C) 2   D) 3

    30.

The end of the contest ✍ **5**

2000-2001 Annual 5th Grade Contest

Spring, 2001

Instructions

5

- **Time** Do *not* open this booklet until you are told by your teacher to begin. You will have only *30 minutes* working time for this contest. You might be *unable* to finish all 30 questions in the time allowed.

- **Scores** Please remember that *this is a contest, not a test*—and there is no "passing" or "failing" score. Few students score as high as 24 points (80% correct). Students with half that, 12 points, *should be commended!*

- **Format and Point Value** This is a multiple-choice contest. Each answer is an A, B, C, or D. Write each answer in the *Answer Column* to the right of each question. A correct answer is worth 1 point. Unanswered questions get no credit. You **may** use a calculator.

43

1. $30 + 40 + 50 = 80 +$?

   A) 50   B) 40   C) 30   D) 20

   1.

2. What number is 50 less than 100 more than 50?

   A) 25   B) 50   C) 75   D) 100

   2.

3. If 2 worms crawl south every hour,
   then ? worms crawl south in 6 hours.

   A) 3   B) 6   C) 8   D) 12

   3.

4. $63 = (6 + 3) \times$?

   A) 7   B) 8   C) 9   D) 10

   4.

5. If every chocolate chip cookie contains 20 chocolate chips, then
   100 chocolate chip cookies contain ? chocolate chips.

   A) 5   B) 120   C) 200   D) 2000

   5.

6. Of the following numbers, which is nearest in value to 111?

   A) 101   B) 109   C) 119   D) 121

   6.

7. An express train travels twice as fast as a local train on the same
   route. If the local train travels the entire route in 6 hours, then
   the express train takes ? hours to travel the entire route.

   A) 3   B) 4   C) 8   D) 12

   7.

8. I got paid $10 every day from June 8
   through June 30, for a total of

   A) $210   B) $220   C) $230   D) $300

   8.

9. $1 \times 2 \times 3 = (10 \times 20 \times 30) \div$?

   A) 6   B) 10   C) 100   D) 1000

   9.

10. Each of the following is a polygon *except* a

    A) circle   B) rectangle   C) square   D) triangle

    10.

11. 11 tens + 11 ones =

    A) 110   B) 111   C) 121   D) 122

    11.

Go on to the next page ▮▮▮➡ **5**

12. 5¢ + 50¢ + $500 = A) $5.55 B) $50.55 C) $500.55 D) $555.00	12.
13. I type 2400 words per hour, so I average _?_ words per minute. A) 24 B) 40 C) 60 D) 120	13.
14. Paul buys a new football every 3 years. Paul bought his first football when he was 8. He bought his fifth football when he was A) 11 B) 19 C) 20 D) 23	14.
15. A rectangle has exactly _?_ pairs of parallel sides. A) none B) 1 C) 2 D) 4	15.
16. The average of all the odd numbers between 2 and 10 is A) 5 B) 6 C) 7 D) 8	16.
17. Which two-digit number is twice the product of its digits? A) 18 B) 26 C) 36 D) 66	17.
18. 20 dimes + 20 nickels = _?_ quarters A) 10 B) 12 C) 20 D) 300	18.
19. The time 10 hours after 10 A.M. is also the time 10 hours before A) 6 A.M. B) 4 A.M. C) 10 P.M. D) 8 P.M.	19.
20. The tens' digit of the product $110 \times 120 \times 130 \times 140 \times 150$ is a A) 0 B) 1 C) 5 D) 6	20.
21. 1 hip = 3 hops and 3 hops = 2 hip-hops, so _?_ hips = 12 hip-hops. A) 4 B) 6 C) 9 D) 18	21.
22. What is the greatest possible remainder when an odd number is divided by 9? A) 0 B) 1 C) 7 D) 8	22.

Go on to the next page ⚡ **5**

45

23. Now, Tina is 10 years older than Ike was 5 years ago. Ike is 15 now. How old will Tina be in 2 years?

A) 27   B) 25   C) 22   D) 20

23.

24. 2 weeks = 7 days + _?_ hours.

A) 24   B) 168   C) 240   D) 336

24.

25. Without folding the paper, I can cut a smaller paper square from a larger one with _?_ (and no fewer) straight cuts with scissors.

A) 1   B) 2   C) 4   D) 8

25.

26. Of the following quotients, which has an odd remainder?

A) $156 \div 12$ B) $259 \div 3$ C) $355 \div 5$ D) $455 \div 3$

26.

27. The product $100 \times 200 \times 300 \times 400$ has _?_ different digits besides 0.

A) 1   B) 2   C) 3   D) 4

27.

28. At a recent *Sing Thing*, it was found that, together, Ann & Bob weigh 180 kg, Carl & Dee weigh 210 kg, and Ann & Carl weigh 220 kg. Together, Bob & Dee weigh

A) 170 kg   B) 180 kg
C) 190 kg   D) 200 kg

28.

29. For how many whole numbers between 100 and 999 does the product of the ones' and tens' digits equal the hundreds' digit?

A) 18   B) 19   C) 21   D) 23

29.

30. Of all rectangles that can be formed from thirty 4×4 squares, the one with the greatest perimeter has a perimeter of

A) 88   B) 136   C) 248   D) 480

30.

The end of the contest 🖎 **5**

Solutions on Page 111 • Answers on Page 147

46

6th Grade Contests

1996-1997 through 2000-2001

1996-97 Annual 6th Grade Contest

Tuesday, March 11, 1997

Instructions

- **Time** You will have only *30 minutes* working time for this contest. You might be *unable* to finish all 40 questions in the time allowed.

- **Scores** Please remember that *this is a contest, not a test*—and there is no "passing" or "failing" score. Few students score as high as 30 points (75% correct). Students with half that, 15 points, *should be commended!*

- **Format and Point Value** This is a multiple-choice contest. Each answer is an A, B, C, or D. Write each answer in the *Answers* column to the right of each question. A correct answer is worth 1 point. Unanswered questions get no credit. You **may** use a calculator.

1. I eat 1 apple a week, on Sunday. In 1 month, I eat at most
   A) 2 apples B) 3 apples C) 4 apples D) 5 apples

1.

2. The average of the first ten odd whole numbers is
   A) 8 B) 9 C) 10 D) 11

2.

3. The number 15 is the greatest common factor of all the following pairs of numbers *except*
   A) 60, 75 B) 75, 90 C) 90, 105 D) 105, 125

3.

4. Of the following, which is *not* equal to the other three?
   A) $32 \div 2$ B) $64 \div 4$ C) $128 \div 6$ D) $192 \div 12$

4.

5. Of the following, the digits of ? have the greatest product.
   A) 1897 B) 1997 C) 2097 D) 2197

5.

6. When a positive number equals its divisor, the quotient must be
   A) 0 B) 1 C) 2 D) 3

6.

7. What is the ones' digit of the product $123\,467\,895 \times 987\,654\,312$?
   A) 1 B) 5 C) 9 D) 0

7.

8. Add 1 to a multiple of 6. Divide the result by 3. The remainder is
   A) 0 B) 1 C) 2 D) 3

8.

9. The sum of six different odd numbers is always
   A) even B) odd C) prime D) less than 90

9.

10. What is the millions' digit of $10^9 + 10^7 + 10^5 + 10^3 + 10^1$?
    A) 0 B) 1 C) 2 D) 3

10.

11. The least positive difference between two unequal primes is
    A) 1 B) 2 C) 3 D) 4

11.

12. All last May, my brother Bobby was scared if and only if the date on the calendar was an odd number. For how many days last May was Bobby scared?
    A) 15 B) 16 C) 17 D) 18

12.

13. ? is *not* a factor of 100010001000.
    A) 3 B) 6 C) 9 D) 12

13.

14. If I had 10 more pennies, I could divide all my pennies into piles of 25 pennies each. The total value of my pennies now could be
    A) $1.35 B) $1.60 C) $1.65 D) $2.10

14.

15. If I need $105, how much must I withdraw from an automatic teller machine that dispenses only $20 bills?
    A) $100 B) $120 C) $125 D) $130

15.

Go on to the next page ⟶ **6**

16. I have \$5. If I spend 20% at the grocery, then I spend 50% of the remaining money at the arcade, how much will I have left? A) \$1.50 B) \$2 C) \$3 D) \$3.50	16.
17. The number of weeds I pulled from my garden was the greatest odd factor of $1 \times 2 \times 3 \times 4 \times 5 \times 6 \times 7 \times 8 \times 9 \times 10$. How many weeds did I pull from my garden? A) 9 B) 567 C) 945 D) 14175	17.
18. The twelfth smallest positive prime is A) 37 B) 31 C) 29 D) 23	18.
19. $(3^3 \div 3) + (3^4 \div 3^2) + (3^5 \div 3^3) = \underline{\ ?\ } \times 3^2$ A) 1 B) 3 C) 3^2 D) 3^3	19.
20. (number of seconds in 1 day) ÷ (number of minutes in 1 day) = A) 24 B) 60 C) 120 D) 3600	20.
21. In a triangle, *no* angle can have a measure of A) 1° B) 120° C) 150° D) 180°	21.
22. A 1×9 rectangle has the same perimeter as a square of area A) 9 B) 20 C) 25 D) 81	22.
23. If 8 ✿ 3 ✿ 4 = 96, which operation is ✿? A) + B) − C) ÷ D) ×	23.
24. In a circle, the ratio of a diameter to a radius is A) 1:2 B) 1:1 C) 2:1 D) 4:1	24.
25. The largest of 10 consecutive whole numbers whose sum is 95 is A) 19 B) 15 C) 14 D) 10	25.
26. If I gave Pat 10 cents, then Pat and I would have the same amount of money. I now have $\underline{\ ?\ }$ cents more than Pat. A) 5 B) 10 C) 15 D) 20	26.
27. The first 3 digits of Pizza Boat's phone number have the same sum as its last 4 digits. If all 7 digits are different, their phone number could be A) 819-4536 B) 319-4503 C) 915-6054 D) 879-6574	27.
28. 100 000 000 000 000 000 000 ÷ $\underline{\ ?\ }$ = 100 000 A) 15 B) 4 C) 10^{15} D) 10^4	28.
29. 1997 minutes is $\underline{\ ?\ }$ minutes more than a whole number of hours. A) 17 B) 5 C) $0.28\overline{3}$ D) $0.208\overline{3}$	29.

Go on to the next page ⟫ **6**

30. My 6 math tests average 12 points more than my 6 art tests. The sum of my math tests is _?_ points more than the sum of my art tests.
    A) 12　　　　　B) 24　　　　　C) 36　　　　　D) 72

30.

31. For every letter, an alphabetized list has 3 words that begin with that letter. The word "SALE" could be the _?_ entry on the list.
    A) 51st　　　　B) 53rd
    C) 54th　　　　D) 56th

31.

32. Each side of a triangle has a whole-number length. If the sum of two sides is 5 cm, then the third side could *not* be _?_ long.
    A) 1 cm　　　　B) 2 cm　　　　C) 3 cm　　　　D) 4 cm

32.

33. The area of each small square is 1. If the two large squares overlap as shown, then the area of the shaded region is
    A) 4　　　B) 4.5　　　C) 5　　　D) 5.5

33.

34. The least whole number whose digits add up to 82 has _?_ digits.
    A) 11　　　　B) 10　　　　C) 9　　　　D) 8

34.

35. If he earned $6 for every crater he stepped in and $7 for every moonrock he picked up, then Astronaut Al could have earned
    A) $22　　B) $23　　C) $29　　D) $32

35.

36. 600% of one hour = _?_ % of one day.
    A) 12　　B) 24　　C) 25　　D) 50

36.

37. $\left(1-\frac{1}{2}\right)\times\left(1-\frac{1}{3}\right)\times\left(1-\frac{1}{4}\right)\times\ldots\times\left(1-\frac{1}{20}\right)=$
    A) $\frac{1}{20}$　　　　B) $\frac{1}{10}$　　　　C) $\frac{9}{10}$　　　　D) $\frac{19}{20}$

37.

38. The sum of 50 consecutive whole numbers equals the sum of 25 other consecutive whole numbers. If the smallest of the 50 numbers is 1997, the smallest of the 25 numbers is
    A) 4032　　　B) 4031　　　C) 3994　　　D) 3993

38.

39. The 80th power of 2 = the _?_ power of 16.
    A) 5th　　B) 10th　　C) 20th　　D) 320th

39.

40. My clock, set correctly at 6 P.M., gained 5 minutes each true hour. When my clock next read 7 A.M., I gave my friend a wake-up call and discovered that the correct time was _?_ A.M.
    A) 5:55　　B) 6　　C) 6:55　　D) 8:05

40.

The end of the contest 🖎 **6**

Solutions on Page 117 • Answers on Page 148

52

1997-98 Annual 6th Grade Contest

Tuesday, March 10, 1998

Instructions

- **Time** You will have only *30 minutes* working time for this contest. You might be *unable* to finish all 40 questions in the time allowed.

- **Scores** Please remember that *this is a contest, not a test*—and there is no "passing" or "failing" score. Few students score as high as 30 points (75% correct). Students with half that, 15 points, *should be commended!*

- **Format and Point Value** This is a multiple-choice contest. Each answer is an A, B, C, or D. Write each answer in the *Answers* column to the right of each question. A correct answer is worth 1 point. Unanswered questions get no credit. You **may** use a calculator.

1. If each car holds 5 clowns, then _?_ cars will hold 60 clowns.
   A) 5   B) 10   C) 12   D) 60

   1.

2. $(5 \times 0) + (6 \times 1) + (7 \times 10) + (8 \times 100) + (9 \times 1000) =$
   A) 98765   B) 56789   C) 9876   D) 6789

   2.

3. How many hours are there in one week?
   A) 7   B) 24   C) 140   D) 168

   3.

4. $1 + 9 + 9 + 8 =$
   A) 2+9+9+8   B) 1+9+9+9   C) 1+8+9+9   D) 1+8+9+8

   4.

5. If you subtract the least positive odd number from the least positive even number, the result will be
   A) 0   B) 1   C) 2   D) 3

   5.

6. 40 and 80 have the same greatest common factor as 80 and
   A) 120   B) 140   C) 160   D) 400

   6.

7. If 130 of 300 envelopes contain letters, how many envelopes do *not* contain letters?
   A) 130   B) 170   C) 270   D) 430

   7.

8. The numbers 18, 27, 36, 45, 54, 63, 72, and 81 are all multiples of
   A) 6   B) 9   C) 12   D) 18

   8.

9. $(3 \times 1998) - 1998 =$
   A) 1×1998 B) 2×1998 C) 3×1998 D) 4×1998

   9.

10. What is three hundred five rounded to the nearest ten?
    A) 300   B) 310   C) 350   D) 400

    10.

11. What is the sum of the measures of all the angles in a square?
    A) 90°   B) 180°   C) 360°   D) 400°

    11.

12. $(61 + 61 + 61 + 39 + 39 + 39) \div (71 + 71 + 71 + 29 + 29 + 29) =$
    A) $1 \div 2$ B) $2 \div 3$ C) $3 \div 4$ D) $5 \div 5$

    12.

13. Together, my three cats weigh 60 kg. If the least weight of any of my cats is 20 kg, then the greatest weight of any of them is _?_ kg.
    A) 20   B) 30   C) 39   D) 40

    13.

14. $(11 + 22 + 33) \div (1 + 2 + 3) =$
    A) 1×11 B) 3×11 C) 6×11 D) 6

    14.

15. Which ratio equals 8:10?
    A) 12:10   B) 10:12   C) 10:8   D) 4:5

    15.

16. How many prime numbers are greater than 0 and less than 40?
    A) 12   B) 13   C) 14   D) 15

    16.

Go on to the next page ⫸ **6**

17. A machine dispenses gumballs always in the order green, blue, yellow, orange, red. If my first gumball was blue, and I bought a total of 8 gumballs, one after the other, then my 8th gumball was A) green　B) blue　C) yellow　D) orange	17.
18. When 3 new kids joined 10 boys and 8 girls, the number of boys increased by 20%. What was the new ratio of boys to girls? A) 4:3　　B) 3:4　　C) 3:2　　D) 2:3	18.
19. Each of the following equals $2^2 + 2^2 + 2^2 + 2^2$ *except* A) $2\times2\times2\times2$　B) 4×2^2　　C) 2^4　　　D) 2×4^2	19.
20. The area of Square A is twice the area of Square B. If the perimeter of Square B is 8 cm, then the area of Square A is A) 64 cm^2　B) 16 cm^2　　C) 8 cm^2　　D) 4 cm^2	20.
21. A log-chewing competition is held once every 4 years. The competition can be held at most _?_ times in any decade. A) 2　　B) 3　　C) 4　　D) 5	21.
22. A telephone pole was placed at the start of a 240 m road and every 12 m thereafter. This road has _?_ telephone poles. A) 10　B) 12　C) 20　D) 21	22.
23. Sniffles sneezed a total of 30 minutes. If each sneeze took 30 seconds, then how many times did Sniffles sneeze? A) 1　　　　B) 10　　　　C) 30　　　　D) 60	23.
24. Five people met. If everyone shook hands once with everyone else, what is the total number of handshakes that occurred? A) 4　　　　B) 5　　　　C) 10　　　　D) 20	24.
25. What is the average of 1, 2, 3, 4, 5, 6, 7, 8, 9, 10, 11, and 12? A) 6.5　　　B) 6　　　　C) 7.5　　　D) 7	25.
26. When ■ = _?_, then ■×■ = ■+■. A) 1　B) 2　C) 3　D) 4	26.
27. Ali and I ran toward each other from 60 m apart. If I ran twice as fast as Ali, I ran _?_ m until we met. A) 10　B) 20　C) 30　D) 40	27.
28. The Movie Theater sells two tickets for the price of one every Monday. The price of one ticket is \$7. If quadruplets go to the movies on Monday, what is the total they must pay for tickets? A) \$7　　　　B) \$14　　　C) \$21　　　D) \$28	28.
29. When juiced, an orange yields 10 ml and a pineapple 30 ml. If 7 oranges and 1 pineapple are juiced, the mixture is _?_ pineapple. A) 25%　　　B) 30%　　　C) 70%　　　D) 75%	29.

Go on to the next page ⇒ **6**

30. Parking costs 50¢ an hour at Pete's, $12 for 24 hours at Paul's, and $75 a week at Patty's. Parking for 14 24-hour days costs at least
    A) $84   B) $140   C) $150   D) $168

30.

31. A side of a square and a radius of a circle are both 4. What is the sum of the areas of the circle and the square?
    A) 20π B) 32π C) $16\pi + 4$ D) $16\pi + 16$

31.

32. A stick 42 cm long is divided into 6 equal parts by _?_ cuts.
    A) 5   B) 6   C) 7   D) 8

32.

33. $(1+2+3+4)^2-(1+2+3+4)\times(1+2+3+4)+(1+2+3+4) =$
    A) 0   B) 10   C) 100   D) 110

33.

34. How many integers between 1 and 1000 are multiples of 17?
    A) 1   B) $1000\div17$ C) 58   D) 59

34.

35. If Bo's score was 8/10 of Al's, and Cy's score was 3/4 of Bo's, then the ratio of Al's score to the average of Bo's and Cy's scores was
    A) 10:7   B) 7:10   C) 5:3   D) 3:5

35.

36. While sitting on an arrow, Pat was catapulted 50 m at a speed of 100 cm/sec. Pat landed in _?_ seconds.
    A) 0.5   B) 2   C) 50   D) 100

36.

37. The product of the first 1000 prime numbers is *not* divisible by
    A) 111   B) 333   C) 555   D) 777

37.

38. Frenzied Frank sells peaches for $1/kg. If Frenzied Frank sold me 15 peaches for $4, the average weight of a peach was _?_ kg.
    A) $\frac{4}{15}$ B) $\frac{15}{4}$ C) $\frac{1}{15}$ D) 1

38.

39. How many whole numbers between 1 and 1000 have 2 as the tens' digit or the hundreds' digit or both digits?
    A) 90   B) 100   C) 190   D) 200

39.

40. When _?_ 1st is a Sunday, then the first days of the next three months are Tuesday, Friday, and Monday, respectively.
    A) June   B) July   C) September   D) October

40.

The end of the contest 🖎 **6**

Solutions on Page 121 • Answers on Page 149

56

1998-99 Annual 6th Grade Contest

Tuesday, March 9, 1999

Instructions

6

- **Time** You will have only *30 minutes* working time for this contest. You might be *unable* to finish all 40 questions in the time allowed.

- **Scores** Please remember that *this is a contest, not a test*—and there is no "passing" or "failing" score. Few students score as high as 30 points (75% correct). Students with half that, 15 points, *should be commended!*

- **Format and Point Value** This is a multiple-choice contest. Each answer is an A, B, C, or D. Write each answer in the *Answers* column to the right of each question. A correct answer is worth 1 point. Unanswered questions get no credit. You **may** use a calculator.

1. Our "Top Banana," who visits only on Fridays, makes at most _?_ such visits in one month. A) 2 B) 3 C) 4 D) 5	1.
2. $(50 \times 60 \times 70 \times 80) \div (5 \times 6 \times 7 \times 8) =$ A) 10 B) 100 C) 1000 D) 10 000	2.
3. All of the following are factors of 10 101 *except* A) 17 B) 13 C) 7 D) 3	3.
4. $99 \times 101 =$ A) 98×102 B) 97×103 C) 100×100 D) 101×99	4.
5. 1 day + 9 hours + 19 minutes = _?_ minutes. A) 29 B) 583 C) 1999 D) 86 959	5.
6. What month comes 4 months *after* the 10th month of the year? A) January B) February C) June D) October	6.
7. The product of the tens' digit and the hundreds' digit of 9876 is A) 72 B) 63 C) 56 D) 54	7.
8. The sum of the first 4 digits on my ticket stub equals the sum of the last 3 digits. My ticket stub number *could* be A) 4 503 219 B) 4 526 819 C) 4 375 891 D) 6 073 915	8.
9. In a square, the ratio of the perimeter to the length of a side is A) 1:4 B) 1:2 C) 2:1 D) 4:1	9.
10. 1999 quarters = _?_ nickels A) $1999 \div 5$ B) 1999×5 C) $1999 + 5$ D) $1999 - 5$	10.
11. How many whole numbers are less than 20 units from 79? A) 19 B) 20 C) 38 D) 39	11.
12. Of the following polygons, which has the most vertices? A) pentagon B) square C) rectangle D) triangle	12.
13. If I exercised on Jan. 1 last year, and on *alternate* days thereafter, I exercised for _?_ days last year. A) 182 B) 183 C) 365 D) 366	13.
14. In the number 12 345, the "1" represents A) 10^2 B) 10^3 C) 10^4 D) 10^5	14.
15. 10 cm + 20 cm + 30 cm + 40 cm = A) 1 m B) 10 m C) 100 m D) 1000 m	15.
16. How many more minutes are there in March than in April? A) 720 B) 1440 C) 2160 D) 2880	16.

Go on to the next page ⇒ **6**

17. A 2×6 rectangle has the same perimeter as a square of area A) 4 B) 12 C) 16 D) 144	17.
18. The sum of two consecutive integers is *always* A) even B) odd C) prime D) at least 10	18.
19. $(19+98) \times (19+99) = 19 \times (19+99) + \underline{\ ?\ } \times (19+99)$. A) 19 B) 97 C) 98 D) 99	19.
20. Mom bought a diving suit for \$50 and sold it for \$100. Based on her cost, she made a profit of $\underline{\ ?\ }$ %. A) 50 B) 100 C) 150 D) 200	20.
21. The sum of the ten smallest positive primes is A) 101 B) 109 C) 110 D) 129	21.
22. If the sum of the digits of a whole number is 200, the whole number may *not* have $\underline{\ ?\ }$ digits. A) 10 000 B) 1000 C) 100 D) 10	22.
23. A rope 2 m long can form a circle with a circumference of A) 2 m B) π m C) 2π m D) 4π m	23.
24. What is the area of the shaded region of the 8 by 12 rectangle shown? A) 24 B) 48 C) 64 D) 96	24.
25. If they were both born on January 1 at noon, a woman in her 90's is *at most* $\underline{\ ?\ }$ years older than an 86-year-old. A) 4 B) 10 C) 13 D) 14	25.
26. The length of a side of a square is an even number. The perimeter of this square *could* be A) 12 B) 30 C) 54 D) 72	26.

27. I hit the panic button every March 9. When I hit it on March 9, 1999, it was a Tuesday. When I next hit it on March 9, 2000, it will be a
   A) Monday   B) Tuesday
   C) Wednesday   D) Thursday
	27.

28. Of the following, which is a multiple of 1999?
   A) $1999^2 + 1998 + 1$ B) $1999^2 + 1$
   C) $1999^2 + 1998$ D) $1999^2 - 1$
	28.

29. A square of side $\underline{\ ?\ }$ and a circle of radius 2 have equal perimeters. A) 1 B) π C) 4 D) 4π	29.

Go on to the next page ⫸ **6**

30. If the sum of all the whole number factors of a number is 1 more than the number, then the number must be
    A) odd   B) even   C) prime   D) composite

30.

31. The product $2 \times 2 \times 2 \times 3 \times 3 \times 3 \times 4 \times 4 \times 4$ equals
    A) $2^8 \times 3^3$ B) $4^2 \times 6^3$ C) $6^3 \times 8^2$ D) $2^4 \times 12^3$

31.

32. How many numbers greater than 1 and less than 500 are multiples of 2 but *not* multiples of 5?
    A) 50   B) 200   C) 249   D) 250

32.

33. The Town of *Us*, with 3400 people, gains 70 people each month. The town of *Them*, with 10 600 people, loses 130 people each month. In how many months will *Us* and *Them* have equal populations?
    A) 36   B) 70   C) 72   D) 120

33.

34. $23 \times 22 \times 21 \times \ldots \times 3 \times 2 \times 1 =$
    A) $27 \times 26 \times 25 \times \ldots \times 10 \times 9 \times 8$ B) $26 \times 25 \times 24 \times \ldots \times 9 \times 8 \times 7$
    C) $25 \times 24 \times 23 \times \ldots \times 8 \times 7 \times 6$ D) $24 \times 23 \times 22 \times \ldots \times 7 \times 6 \times 5$

34.

35. My class was lined up on the gym floor in 9 rows, with 4 students in each row. If our coach rearranged us so that the number of rows was the same as the number of students in each row, how many rows were there after we were rearranged?
    A) 8   B) 6   C) 5   D) 4

35.

36. The least common multiple of 51 and 117 is
    A) 1   B) 117×3 C) 117×17 D) 117×51

36.

37. When Pat wrote 100 numbers, each less than 100, on a notepad, Pat's magic pencil subtracted each number from 100, then wrote all 100 results on the notepad. What is the sum of all 200 numbers Pat and the Magic Pencil wrote on the notepad?
    A) 10 000   B) 20 000   C) 30 000   D) 40 000

37.

38. The 160th power of 2 equals the _?_ power of 32.
    A) 5th   B) 10th   C) 25th   D) 32nd

38.

39. If all angles of a triangle are perfect squares, its smallest angle is
    A) 4°   B) 9°   C) 16°   D) 36°

39.

40. Each number after the first two numbers in a sequence is the sum of all the previous numbers. If the 10th number is 1000, then the 8th number is
    A) 250   B) 500   C) 800   D) 1000

40.

The end of the contest 🖝 **6**

Solutions on Page 125 • Answers on Page 150

1999-2000 Annual 6th Grade Contest

Tuesday, March 14, 2000

Instructions

6

- **Time** You will have only *30 minutes* working time for this contest. You might be *unable* to finish all 40 questions in the time allowed.

- **Scores** Please remember that *this is a contest, not a test*—and there is no "passing" or "failing" score. Few students score as high as 30 points (75% correct). Students with half that, 15 points, *should be commended!*

- **Format and Point Value** This is a multiple-choice contest. Each answer is an A, B, C, or D. Write each answer in the *Answers* column to the right of each question. A correct answer is worth 1 point. Unanswered questions get no credit. You **may** use a calculator.

1. The sum $1 + 2 + 34 + 56$ equals all of the following *except* A) $36 + 57$ B) $37 + 56$ C) $35 + 58$ D) $37 + 59$	1.
2. Divide $4 \times 5 \times 6 \times 7 \times 8 \times 9 \times 10 \times 11 \times 13$ by 7. The remainder is A) 0 B) 3 C) 4 D) 6	2.
3. Of 27 mice, 24 are not hurrying. How many are hurrying? A) 2 B) 3 C) 12 D) 24	3.
4. 11 ones + 11 tens + 11 hundreds = A) 111 111 B) 12 321 C) 1221 D) 144	4.
5. $2 \times 4 \times 6 \times 8 \times 10 \div$? $= 1 \times 2 \times 3 \times 4 \times 5$ A) 2 B) 5 C) 16 D) 32	5.
6. Which of the following is twice as large as $36 + 64$? A) $2 \times 36 + 64$ B) $2 + 36 + 64$ C) $36 + 64 \times 2$ D) $72 + 128$	6.
7. Every even whole number is divisible by A) 1, but not 0 B) 0, but not 1 C) 0 and 1 D) 0, 1, and 2	7.
8. My bird sings 3 hours every day. In 6 weeks, it sings a total of A) 9 hours B) 18 hours C) 42 hours D) 126 hours	8.
9. Which does *not* equal 50 when rounded to the nearest 10? A) 45.45 B) 50.50 C) 52.52 D) 55.55	9.
10. What is the smallest possible difference between two primes? A) 1 B) 2 C) 3 D) 4	10.
11. If 8 kids share the cost of a present equally, each pays $1.20. If only 6 kids share the cost instead, each share would increase by A) 15¢ B) 20¢ C) 30¢ D) 40¢	11.
12. $230\,000\,000 - 20\,000\,000 =$? million A) 3 B) 21 C) 210 D) 228	12.
13. If 1 dinosaur hatches every 12 seconds, then ? dinosaurs hatch every half-hour. A) 150 B) 216 C) 240 D) 360	13.
14. The number "0" is A) even B) odd C) prime D) negative	14.
15. What is the average of the measures of the angles in a triangle? A) 45° B) 60° C) 90° D) 180°	15.
16. Which of the following is 2 less than 4 more than half itself? A) 1 B) 2 C) 4 D) 8	16.

Go on to the next page ⟩⟩⟩➡ **6**

62

17. What is the largest even factor of $6 \times 7 \times 8 \times 9 \times 10$?
   A) 32   B) 160   C) 480   D) 30 240

17.

18. If a teeter-totter teeters once every 10 totters, how many times does it teeter for every 200 totters?
   A) 2000   B) 210   C) 20   D) 10

18.

19. Starting at noon, I watched TV for 37 hours, then stopped. At what time did I stop?
   A) 11 A.M.   B) 1 A.M.
   C) 12 P.M.   D) 1 P.M.

19.

20. If 9 of 30 students got A's, what percent *didn't* get A's?
   A) 21%   B) 65%   C) 70%   D) 79%

20.

21. Each of 5 kids played checkers with each of the other 4 once a day. After some whole number of days of doing this, the total number of such games played by one of these kids could *not* be
   A) 8   B) 12   C) 18   D) 20

21.

22. On a number line, ? is the same distance from 0.25 as it is from 4.
   A) 1.875   B) 2.0625   C) 2.125   D) 2.25

22.

23. Cathy sells cakes for $9 each. If it costs Cathy $10 to make 4 cakes, how much profit does Cathy make on each cake sold?
   A) $0.25   B) $1.00   C) $2.50   D) $6.50

23.

24. Whether I double or square the number ? , I get the same result.
   A) 1   B) 2   C) 4   D) 16

24.

25. Since I only dance on alternate days, next week I'll dance at most ? days.
   A) 2   B) 3   C) 4   D) 5

25.

26. Ann is 2 cm shorter than Bob, who is 1 cm shorter than Carl. If Dee is 4 cm taller than Ann, who is tallest?
   A) Ann   B) Bob   C) Carl   D) Dee

26.

27. When I multiply the lengths of all 4 sides of a square, I get 256. When I add the lengths of all 4 sides, I get
   A) 16   B) 32   C) 64   D) 256

27.

28. Together, 9 dimes and ? nickels are worth as much as 6 quarters.
   A) 3   B) 12   C) 15   D) 30

28.

29. Which is *not* the sum of two different whole numbers?
   A) 0   B) 1   C) 2   D) 100

29.

Go on to the next page ⫸ **6**

30. The greatest common factor of 6^3 and 4^5 is A) 8 B) 12 C) 16 D) 136	30.
31. From a mix of 12 white, 12 black, and 12 blue socks, I must take at least ? socks to guarantee I have at least one matching pair. A) 3 B) 4 C) 13 D) 14	31.
32. I spent \$12 on pens that cost 40¢ each and another \$12 on pens that cost 60¢ each. Altogether, my average cost per pen was A) 45¢ B) 48¢ C) 50¢ D) 52¢	32.
33. $2^3 \times 3^3 \times 4^3 \times 6^3 \times 9^3 =$ A) $2^9 \times 3^9$ B) $2^{12} \times 3^{12}$ C) $2^{15} \times 3^{15}$ D) $2^{54} \times 3^{54}$	33.
34. In a race, if my time beat your time by 12 fewer seconds than my time beat Pat's time, then your time beat Pat's time by A) 6 seconds B) 12 seconds C) 18 seconds D) 24 seconds	34.
35. If two positive integers have a sum of 12, their ratio *cannot* be A) 1:2 B) 1:3 C) 1:4 D) 1:5	35.
36. The sum of the first 10 000 positive even numbers is ? more than the sum of the first 10 000 positive odd numbers. A) 1 B) 5000 C) 9999 D) 10 000	36.
37. If the area of the shaded part of the square is 16, then the perimeter of the square is A) 8 B) 16 C) 32 D) 64	37.
38. Divide 2000 by an odd number. The remainder *must* be A) even B) odd C) prime D) whole	38.
39. $2^{2001} - 2^{2000} - 2^{1999} - 2^{1998} =$ A) 2 B) 2^4 C) 2^{500} D) 2^{1998}	39.
40. Lois and Clark had a foot race. Clark had a headstart of 5 seconds. If Lois caught up to Clark after running 60 m in 10 seconds, then Clark's average running speed was A) 2 m/sec B) 4 m/sec C) 6 m/sec D) 14 m/sec	40.

The end of the contest ✍ **6**

Solutions on Page 129 • Answers on Page 151

2000-2001 Annual 6th Grade Contest

Tuesday, March 13 or 20, 2001

Instructions

6

- **Time** You will have only *30 minutes* working time for this contest. You might be *unable* to finish all 40 questions in the time allowed.

- **Scores** Please remember that *this is a contest, not a test*—and there is no "passing" or "failing" score. Few students score as high as 30 points (75% correct). Students with half that, 15 points, *should be commended!*

- **Format and Point Value** This is a multiple-choice contest. Each answer is an A, B, C, or D. Write each answer in the *Answers* column to the right of each question. A correct answer is worth 1 point. Unanswered questions get no credit. You **may** use a calculator.

1. $77 - (33 + 44) = (77 - 33) - \underline{\ ?\ }$
   A) 44   B) 33   C) 22   D) 0

   1.

2. Kat the lion tamer tames 9 lions every year. How many lions does Kat tame in 9 years?
   A) 18   B) 81   C) 90   D) 99

   2.

3. If I divide my house number by 2 and then add 6, I get 24. What is my house number?
   A) 9   B) 15   C) 36   D) 54

   3.

4. In which of the following divisions will the remainder equal 2?
   A) $72 \div 36$ B) $218 \div 18$ C) $416 \div 16$ D) $616 \div 14$

   4.

5. If I subtract the number which is 1 less than 1 million from the number which is 1 more than 1 million, then the difference is
   A) 1   B) 2   C) 1 000 000   D) 1 000 001

   5.

6. The average age of 5 bears is 120 months. The sum of their ages is $\underline{\ ?\ }$ years.
   A) 10   B) 24   C) 50   D) 120

   6.

7. The number $\underline{\ ?\ }$ is divisible by 3×3.
   A) 663   B) 603   C) 336   D) 303

   7.

8. (# of days in a year) − (# of days in February of that year) =
   A) 335   B) 336   C) 337   D) 338

   8.

9. If a typical gorilla eats 2 dozen bananas daily, how many bananas do 10 typical gorillas eat daily?
   A) 12   B) 20   C) 200   D) 240

   9.

10. The tens' digit of the product $111\,111\,111\,111 \times 111\,111\,111\,111$ is
    A) 0   B) 1   C) 2   D) 3

    10.

11. Every prime has exactly $\underline{\ ?\ }$ different whole number factors.
    A) 0   B) 1   C) 2   D) 3

    11.

12. If a helicopter can fly for 90 minutes on 1 full tank of gas, how many full tanks of gas does it need to fly for 6 hours?
    A) 3   B) 4   C) 15   D) 60

    12.

13. *Between* the 1st person and the 51st person on a line, there are $\underline{\ ?\ }$ people.
    A) 49   B) 50   C) 51   D) 52

    13.

14. $2^2 \times 2^3 \times 2^4 \times 2^5 =$
    A) 60   B) 2^6 C) 240   D) 4^7

    14.

15. Divide an odd number by 4. The remainder is *always*
    A) odd   B) even   C) 1   D) prime

    15.

Go on to the next page ⟩⟩⟩ **6**

16. What is the greatest common factor of 8^2 and 2^8?

A) 2^2 B) 2^3 C) 4^2 D) 4^3

16.

17. When 249 973 is rounded to the nearest thousand, how many of its digits are not changed?

A) 1   B) 2   C) 3   D) 5

17.

18. 300 g is _?_ % of 6 kg.

A) 2   B) 5   C) 50   D) 5000

18.

19. Today, 8 kids are taking turns diving, always in the same order. If 2 kids have dived since Dave dove, how many other kids must dive before Dave dives again?

A) 3   B) 4   C) 5   D) 6

19.

20. $\sqrt{100-64} + \sqrt{25-16} =$

A) $\sqrt{81}$ B) $\sqrt{45}$ C) $\sqrt{9}$ D) -65

20.

21. What is the greatest prime factor of $6^3 \times 18^3 \times 30^3$?

A) 2   B) 3   C) 5   D) 6^3

21.

22. What is the square root of the number whose square is 16?

A) 2   B) 4   C) $\sqrt{8}$ D) 16

22.

23. A leap year exceeds 52 weeks by

A) 1 day   B) 2 days   C) 3 days   D) 1 week

23.

24. If one angle of a triangle is 40°, the other angles average

A) 70°   B) 140°   C) 160°   D) 180°

24.

25. If I rode my bike 12 km in 1 hour, then I'll average _?_ m/minute.

A) 20   B) 72   C) 200   D) 720

25.

26. No _?_ number is ever divisible by 2.

A) even   B) prime   C) whole   D) odd

26.

27. If a circle needs 8 360° rotations to roll 8 m, its radius is _?_ m long.

A) 1   B) $\pi \div 2$ C) 2   D) $1 \div (2\pi)$

27.

28. My watch gained 3 minutes the 1st hour, 6 minutes the 2nd hour, 9 minutes the 3rd hour, and so on. In 10 hours, it gained a total of

A) 30 minutes   B) 135 minutes   C) 165 minutes   D) 198 minutes

28.

29. The product of all 4 sides of a square is 1296. The sum of all 4 is

A) 24   B) 36   C) 48   D) 72

29.

30. I am thinking of a whole number greater than 0 whose square equals its square root. How many such numbers are there?

A) 0   B) 1   C) 2   D) 4

30.

Go on to the next page ⏩ **6**

31.	On her birthday today, Ali's age in months is twice her age in years 60 *years from now*. How old is Ali now, in months? A) 12 months B) 24 months C) 120 months D) 144 months	31.
32.	Britney spears her grilled shrimp so the product of the numbers of shrimp on Britney's spears is 1001. If the number of shrimp on each spear is a prime, how many shrimp does Britney spear altogether? A) 31 B) 37 C) 103 D) 151	32.
33.	I ran 1 km in 6.0 minutes. Rabbit ran 50% faster and finished in A) 3.0 minutes B) 4.0 minutes C) 4.5 minutes D) 9.0 minutes	33.
34.	The sum of 10 consecutive integers, starting with 11, equals the sum of 5 consecutive integers, starting with A) 22 B) 29 C) 31 D) 33	34.
35.	If the quotient equals the divisor, then the dividend equals the A) $\sqrt{\text{divisor}}$ B) divisor C) divisor^2 D) quotient	35.
36.	Of the following, which reads the same forwards, backwards, and upside down (when viewed on a calculator screen)? A) 6889 B) 1991 C) 1961 D) 1881	36.
37.	(sum of 50 different whole numbers, each less than 51) − (sum of 45 different whole numbers, each less than 51) is *at most* A) 10 B) 15 C) 240 D) 285	37.
38.	A list of 300 numbers starts at 1. After that, every number is triple the number which precedes it. The 200th number on the list is A) 600 B) 900 C) 3^{199} D) 3^{200}	38.
39.	My piggy bank has only nickels, dimes, and dollar bills. The ratio of nickels to dimes is 2:3, and the ratio of dimes to dollar bills is 8:1. What is the ratio of coins to dollar bills? A) 40:3 B) 16:3 C) 13:1 D) 5:1	39.
40.	Subtract the product of the first 19 positive whole numbers from the product of the first 20 positive whole numbers. The result is A) 19 B) $19 \times 18 \times 17 \times \ldots \times 3 \times 2 \times 1$ C) 20 D) $19 \times 19 \times 18 \times 17 \times \ldots \times 3 \times 2 \times 1$	40.

The end of the contest ✍ **6**

Solutions on Page 133 • Answers on Page 152

Detailed Solutions

• • • • • • • • • • • • • • • • •

1996-1997 through 2000-2001

4th Grade Solutions

1996-1997 through 2000-2001

Information & Solutions

Spring, 1997

Contest Information

4

- **Solutions** Turn the page for detailed contest solutions (written in the question boxes) and letter answers (written in the *Answer Column* to the right of each question).

- **Scores** Please remember that *this is a contest, not a test*—and there is no "passing" or "failing" score. Few students score as high as 24 points (80% correct). Students with half that, 12 points, *deserve commendation!*

- **Answers & Rating Scale** Turn to page 138 for the letter answers to each question and the rating scale for this contest.

1. $(2 + 8) + 10 = (10) + 10 = 2 \times 10.$
   A) 8   B) 10   C) 12   D) 18

   1. B

2. The largest sum is the one without a 0.
   A) 1110+9990   B) 1101+9909
   C) 1011+9099   D) 1111+9999

   2. D

3. The greatest whole number less than 100 is 99, and 10 more than that is 109.
   A) 99   B) 100   C) 109   D) 110

   3. C

4. The sum $50 + 50 + 50 = 150$. That is *not* the value of choice B.
   A) 75+75 = 150   B) 35+35+35+35 = 140
   C) 30+30+30+30+30 = 150   D) 25+25+25+25+25+25 = 150

   4. B

5. 3 dozen = $3 \times 12 = 36$; so 3 more than that = 39.
   A) 24   B) 27   C) 36   D) 39

   5. D

6. $444+444+444 = 400+44 + 400+44 + 400+44 = (3 \times 400)+(3 \times 44).$
   A) 38   B) 40   C) 42   D) 44

   6. D

7. I had exactly 18 lollipops. If I gave away 4, lost 2, and ate 5, the number of lollipops I'd have left would be $18-4-2-5 = 7$.
   A) 7   B) 9   C) 11   D) 18

   7. A

8. All 10 days are checked off in the picture!
   A) Friday   B) Saturday
   C) Sunday   D) Monday

   8. A

9. 80 is a multiple of 8. Start adding 8's to get 88, 96, 104, ... The largest < 100 is 96.
   A) 88   B) 96   C) 98   D) 104

   9. B

10. 10-19, 20-29, 30-39, 40-49 have ten 2-digit #'s each.
    A) 50   B) 49   C) 40   D) 39

    10. C

11. (Odd+odd) and (even+even) are even. (Even+odd) is odd.
    A) 676, 989   B) 687, 989   C) 766, 898   D) 898, 988

    11. A

12. A pizza pie is cut into 8 slices. If each slice is cut into 3 pieces, then there are $8 \times 3 = 24$ pieces altogether.
    A) 11   B) 16   C) 24   D) 38

    12. C

Go on to the next page ▪▪▪▶ 4

74

1996-97 4TH GRADE CONTEST SOLUTIONS

13. If 1 m = 100 cm, then 12 m = 12×1 m = 12×100 cm = 1200 cm. A) 12 cm B) 120 cm C) 1200 cm D) 12 000 cm	13. C
14. 5 quarters are worth $1.25 = $1.20 + 5¢ = 12 dimes + 1 nickel. A) 6 B) 10 C) 12 D) 13	14. C
15. Each seat on my school bus holds 2 students. If my bus has 18 seats, then at most 18×2 = 36 students can sit on my bus. A) 9 B) 16 C) 20 D) 36	15. D
16. If I'm 10 now, 2 years ago I was 8, and 6 years from then I'd be 14. That's in 4 years. A) 4 B) 6 C) 8 D) 12	16. A
17. Every whole number is divisible by 1. A) 3 B) 2 C) 1 D) 0	17. C
18. Both (even − odd) and (odd − even) = odd. Example: 10−1 = 9. A) prime B) 1 C) even D) odd	18. D
19. Doubling gives you an even number. Subtract 2 to get another even number, which is divisible by 2. Example: (2×8) − 2 = 14. A) 2 B) 3 C) 4 D) 5	19. A
20. Since a square has 4 sides and a triangle has 3 sides, a square has 1 more side than a triangle. A) square, triangle B) square, rectangle C) triangle, square D) rectangle, square	20. A
21. If Ali will be 10 in 1998, she'll be 15 in 1998+5 = 2003. A) 2002 B) 2003 C) 2012 D) 2013	21. B
22. You spent $6.25 on lunch and I spent 50¢ less. On lunch, I spent $6.25 − $0.50 = $5.75. A) $1.25 B) $5.75 C) $5.85 D) $6.75	22. B
23. If one coin is worth twice the other, the two coins might be a dime and a nickel, worth 15¢. A) 6¢ B) 15¢ C) 30¢ D) 35¢	23. B

Go on to the next page ⅢⅢ➡ **4**

24. In HiWay City, odd-numbered routes run North and South, and even-numbered routes run East and West. I can travel East on Route 66, Route 70, and Route 98, but not on Route 89.

    Route 66 Route 70 Route 89 Route 98

    A) 1   B) 2   C) 3   D) 4

24.

C

25. He painted 1 face yesterday. Today he paints 2 faces; 1 day from now, 4 faces; 2 days from now, 8 faces; 3 days from now, 16 faces; 4 days from now, 32 faces.

    A) 4   B) 8   C) 16   D) 32

25.

D

26. Choose 3 *different* numbers. Choose a 7 from (6,7,8), an 8 from (2,5,8), and a 6 from (4,6,8) for a sum of 21. Don't choose more than one 8 (the numbers must be *different*).

    A) 24   B) 21   C) 20   D) 19

26.

B

27. Multiply only the last 3 digits of each: $999 \times 888 = 887\underline{1}12$.

    A) 1   B) 2   C) 4   D) 9

27.

A

28. A square whose side is 12 cm long can be cut up into 9 squares whose sides are each 4 cm long, as shown.

    A) 3   B) 6   C) 9   D) 12

28.

C

29. This year, the groundhog saw its shadow 120 times. Last year, it saw its shadow 6 times for every 5 times it saw its shadow this year. There are 24 groups of 5 in 120, so the groundhog saw its shadow $6 \times 24 = 144$ times last year.

    A) 100   B) 144   C) 220   D) 264

29.

B

30. Regroup: $(2000-1000)+(1999-999)$ $+...+(1002-2)+(1001-1)+1000 =$ $1000 \times 1000 + 1000 = 1000 \times 1001$.

    A) 999   B) 1000
    C) 1001   D) 1002

30.

C

The end of the contest 🖎 **4**

Visit our web site at http://www.mathleague.com

Information & Solutions

Spring, 1998
Contest Information

4

- **Solutions** Turn the page for detailed contest solutions (written in the question boxes) and letter answers (written in the *Answer Column* to the right of each question).

- **Scores** Please remember that *this is a contest, not a test*—and there is no "passing" or "failing" score. Few students score as high as 24 points (80% correct). Students with half that, 12 points, *deserve commendation!*

- **Answers & Rating Scale** Turn to page 139 for the letter answers to each question and the rating scale for this contest.

1. The sum of 500 ones is $500 \times 1 = 500$.

   A) 1   B) 500   C) 501   D) 5000

2. 10 hours $= 10 \times 60$ minutes $= 600$ minutes.

   A) 6   B) 60   C) 70   D) 600

3. $19 + 98 = (19+1) + (98-1) = 20 + 97$.

   A) 78   B) 97   C) 99   D) 1978

4. $16 \times 16 = 256$.

   A) 16   B) 32   C) 64   D) 128

5. The number two thousand one is written 2001. It has 4 digits.

   A) 2   B) 3   C) 4   D) 5

6. Since $24 = 3 \times 8$, 24 people can be split into 8 groups of 3 people each.

   A) 4   B) 6   C) 8   D) 12

7. 30 ones $= 30 = 3 \times 10 = 3$ tens.

   A) 3   B) 10   C) 30   D) 300

8. $(12 \div 4) \times 3 = 3 \times 3 = 9$.

   A) 1   B) 3   C) 6   D) 9

9. $(10 \times 1) + (10 \times 10) + (10 \times 100) = 10 + 100 + 1000 = 1110$.

   A) 111   B) 1100   C) 1110   D) 1111

10. 12 *pairs* of red socks is the same as $12 \times 2 = 24$ red socks.

    A) 6   B) 12   C) 14   D) 24

11. There are $125 \div 5 = 25$ nickels in \$1.25. There are $125 \div 25 = 5$ quarters in \$1.25. There are $25 - 5 = 20$ more nickels than quarters.

    A) 20   B) 21   C) 25   D) 30

12. The number 6 is an even number. Since $18 \div 6 = 3$, 6 is a factor of 18.

    A) 9   B) 6   C) 4   D) 3

Go on to the next page ⫸ **4**

13. *Groundbreaking* has vowels *o*, *u*, *e*, *a*, and *i*.

A) 2   B) 3   C) 4   D) 5

13.
D

14. $(2\times1)+(2\times3)+(2\times5) = 2\times(1+3+5) = 2\times9.$

A) 2×8 B) 2×9 C) 6×9 D) $2+9$

14.
B

15. (70×100) cm $= 70\times(100$ cm$) = 70$ m.

A) 7   B) 70   C) 700   D) 7000

15.
B

16. $5555 - 1234 = 4321.$

A) 4021   B) 4123   C) 4231   D) 4321

16.
D

17. I'm next to last of 8 people
holding a hot dog, so $8-2 =$
6 people are ahead of me.

A) 4   B) 5   C) 6   D) 7

17.
C

18. The days are Tuesday, Wednesday, Thursday, and Saturday.

A) 6   B) 5   C) 4   D) 3

18.
C

19. If I double 32, I get 64. Then, $64 \div 4 = 16.$

A) 2   B) 32   C) 64   D) 128

19.
B

20. I misspelled 5 words on a quiz. I spelled 5 times as many, 25,
correctly. Altogether there were $5+25 = 30$ words on the quiz.

A) 5   B) 10   C) 25   D) 30

20.
D

21. If a jigsaw puzzle has 500 pieces,
with 85 edge pieces, then $500-85 =$
415 of the pieces are *not* edge pieces.

A) 315   B) 415   C) 425   D) 585

21.
B

22. Two (3×333)'s $= 2\times3\times333 = 3\times666.$

A) 111   B) 333   C) 666   D) 999

22.
C

23. 1998 is 3 more than 1995, so $1998 \div 5$ leaves a remainder of 3.

A) 1998   B) 1999   C) 2001   D) 2002

23.
A

Go on to the next page ⮕ **4**

24. Tim turned 10 years old exactly two months ago. Tim will turn 12 years old in 24 − 2 = 22 months.

    A) 20   B) 22   C) 24   D) 26

    24.

    B

25. A number divisible by 8 (such as 8) may not be divisible by 6, but is divisible by 1, 2, and 4.

    A) 6   B) 4   C) 2   D) 1

    25.

    A

26. 891 ÷ 3 = 297. Since 297 = 27 × 11, 297 is divisible by 27.

    A) 7   B) 17   C) 27   D) 97

    26.

    C

27. Pat's pot-bellied pig eats 3 pans of pig food a day at a cost of 75¢ per pan. The cost each week for pig food is 75¢ × 3 × 7 = $15.75.

    A) $15.75   B) $5.25   C) $2.25   D) $21

    27.

    A

28. Multiply 2's until the product exceeds 1000: 2×2×2×2×2×2×2×2×2×2 = 1024, so the fewest number of 2's needed is 10.

    A) 4   B) 9   C) 10   D) 501

    28.

    C

29. Average of 1 to 31 is 16, and 16 × 31 = 496. Alternatively, the sum of the 16 odd numbers is even. The sum of the 15 even numbers is also even. The total is even.

    A) 30   B) 31   C) 465   D) 496

    29.

    D

30. To form the number 13 471 897, begin with 1, then 3. Each following digit is the ones' digit of the sum of the two digits before it. The 25-digit number formed the same way, but starting with 1, then 5, is 1 561 785 381 909 987 527 965 167.

    A) 1   B) 5   C) 7   D) 9

    30.

    C

The end of the contest ☞ **4**

Visit our web site at http://www.mathleague.com

Information & Solutions

Spring, 1999

Contest Information

4

- **Solutions** Turn the page for detailed contest solutions (written in the question boxes) and letter answers (written in the *Answer Column* to the right of each question).

- **Scores** Please remember that *this is a contest, not a test*—and there is no "passing" or "failing" score. Few students score as high as 24 points (80% correct). Students with half that, 12 points, *deserve commendation!*

- **Answers & Rating Scale** Turn to page 140 for the letter answers to each question and the rating scale for this contest.

1. Whenever 1's are multiplied and divided, the result is always 1.
   A) 0   B) 1   C) 9   D) 10

   1.
   B

2. To eat here, you need a fork, a spoon, and a knife. Each of the 5 of us needs 3 utensils. All together, we need a total of $5 \times 3 = 15$ utensils.
   A) 5   B) 8   C) 10   D) 15

   2.
   D

3. $(1 \times 2 \times 3 \times 4 \times 5) \div 6 = 120 \div 6 = 20$.
   A) 6   B) 20   C) 36   D) 720

   3.
   B

4. Last month, my parrot ate 16 oranges, 12 bananas, and 22 apples. That was $16+12+22 = 50$ pieces of fruit.
   A) 40   B) 46   C) 48   D) 50

   4.
   D

5. For 1999, hundreds' digit + ones' digit = 9 + 9 = 18.
   A) 10   B) 18   C) 28   D) 81

   5.
   B

6. Add, subtract, or multiply whole numbers: result is a whole number.
   A) $3 \div 2$ B) $3 - 2$ C) 3×2 D) $3 + 2$

   6.
   A

7. The months Feb., Sept., and Dec. have an e as their second letter.
   A) 1   B) 2   C) 3   D) 4

   7.
   C

8. I ate 4 of my two dozen donuts, so I had $24 - 4 = 20$ left.
   A) 8   B) 16   C) 20   D) 28

   8.
   C

9. $32 \div 4 = 8 = 64 \div 8$.
   A) 2   B) 6   C) 8   D) 16

   9.
   C

10. If Michael Jordan's two uniform numbers are 23 and 45, their sum is $23 + 45 = 68$.
    A) 22   B) 68   C) 72   D) 1035

    10.
    B

11. If you multiply by 0, the product is 0.
    A) 0   B) 1   C) 20   D) 384

    11.
    A

12. 2 P.M. + 61 mins = 2 P.M. + 1 hr + 1 min = 3 P.M. + 1 min.
    A) 2:59 P.M.   B) 2:61 P.M.   C) 3:01 P.M.   D) 3:31 P.M.

    12.
    C

Go on to the next page ⫸ **4**

13. Since 6 of the 9 letters in *consonant* are consonants, and 2 of the 5 letters in *vowel* are vowels, the sum is 6 + 2 = 8. A) 3 B) 8 C) 9 D) 11	13. B
14. Seven pennies + odd number of nickels has ones' digit of 2. Seven pennies + even number of nickels has ones' digit of 7. Hence, ones' digit must be a 2 or a 7. A) $66.66 B) $67.67 C) $77.77 D) $222.22	14. A
15. $(2 \times 3 \times 4 \times 5 \times 6) \div (2+3+4+5+6) = 36$. A) 12 B) 24 C) 36 D) 60	15. C
16. 33 less than 44 = 44−33 = 11; 22 more than that is 22+11 = 33. A) 99 B) 55 C) 33 D) 11	16. C
17. The product is even *except* when both numbers you multiply are odd. A) 11 × 99 B) 44 × 33 C) 55 × 22 D) 88 × 66	17. A
18. 50¢ per pack is the same as 2 for $1. Thus, for $5 you can buy 2 × 5 = 10 packs of baseball cards. A) 5 B) 10 C) 20 D) 50	18. B
19. I am 9. My brother is 2 × 9 = 18. The referee is 25 years older than my brother, so the ref is 25+18 = 43. A) 43 B) 48 C) 50 D) 59	19. A
20. The total number of 7's is 8. A) 7 B) 8 C) 49 D) 56	20. B
21. To compute the *square* of a number, just multiply the number by itself. So 11 × 11 = 121. A) 22 B) 110 C) 111 D) 121	21. D
22. Subtract 10 from 11, 20 from 22, ..., 80 from 88; 10+20+...+80 = 360. A) 10 B) 80 C) 180 D) 360	22. D

Go on to the next page ⦙⦙⮕ **4**

23. One million = 1 000 000. Adding, $1+0+0+0+0+0+0 = 1$. A) one B) one hundred C) one thousand D) one million	23. A
24. Alex is 4 years younger than Lee. That includes a leap year, so Alex is $365+365+365+366 = 1461$ days younger than Lee. A) 365 B) 1460 C) 1461 D) 1464	24. C
25. For a party, we ordered four pizzas shaped like a square (4 sides) and one shaped like a pentagon (5 sides). When added together, the total number of sides was $4+4+4+4+5 = 21$. A) 17 B) 18 C) 20 D) 21	25. D
26. 7 years from now (8 years from last year), I'll be twice as old as I was last year. Thus, I was 8 last year; and I'm 9 this year. A) 9 B) 8 C) 7 D) 6	26. A
27. If 6 divides evenly into my age and my grandmother's age, then the sum of our ages also has to be evenly divisible by 6. A) 52 B) 54 C) 56 D) 58	27. B
28. See the picture. $\overline{BD}$ would be a diagonal of $ABCD$. B C A) $\overline{BD}$ B) $\overline{AD}$ C) $\overline{CD}$ D) $\overline{AB}$ A D	28. A
29. We brought in 5 cans for every 3 cans you brought in. You brought in 60 cans, or 3 groups of 20. We brought in 5 groups of 20, a total of 100 cans. A) 36 B) 65 C) 75 D) 100	29. D
30. Of the whole numbers from 1 through 100, the numbers $1, 2, \ldots, 90$ are each 5 less than another number from 1 through 95. A) 20 B) 90 C) 95 D) 100	30. B

The end of the contest 🖐 **4**

Visit our web site at http://www.mathleague.com

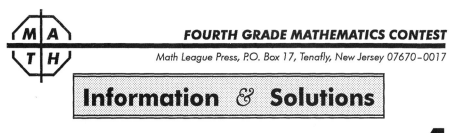

Information & Solutions

Spring, 2000

Contest Information

4

- **Solutions** Turn the page for detailed contest solutions (written in the question boxes) and letter answers (written in the *Answer Column* to the right of each question).

- **Scores** Please remember that *this is a contest, not a test*—and there is no "passing" or "failing" score. Few students score as high as 24 points (80% correct). Students with half that, 12 points, *deserve commendation!*

- **Answers & Rating Scale** Turn to page 141 for the letter answers to each question and the rating scale for this contest.

1. $999 + 999 + 2 = (999 + 1) + (999 + 1) = 1000 + 1000 = 2000$.
   A) 1001   B) 1998   C) 2000   D) 2002

   1. C

2. 2 more than 2 dozen mints = 2 dozen + 2 = 24 + 2 = 26 mints.
   A) 14   B) 22   C) 26   D) 48

   2. C

3. Every day, my elephant begs for 3 bags of peanuts. Each week it begs for $7 \times 3 = 21$ bags.
   A) 3   B) 7   C) 10   D) 21

   3. D

4. $50 + 100 + 150 = 300 = 6 \times 50$.
   A) 2   B) 3   C) 4   D) 6

   4. D

5. The hundreds' digit of choice A is a 7 and the tens' digit is a 6, so choice A is the answer.
   A) 9764   B) 8459   C) 1234   D) 1000

   5. A

6. $19 + 91 = 110$, while $18 + 81 = 27 + 72 = 36 + 63 = 99$.
   A) 19 + 91   B) 18 + 81   C) 27 + 72   D) 36 + 63

   6. A

7. 98 766 789 is the same forwards and backwards (it's a *palindrome*).
   A) 98 766 789   B) 45 545 454   C) 12 343 214   D) 10 535 301

   7. A

8. (Number of sides in a triangle) × (number of sides in a square) $= 3 \times 4 = 12$, and 5 is *not* a factor of 12.
   A) 3   B) 4   C) 5   D) 6

   8. C

9. I ate 3 times as many ice cream bars as frozen yogurt bars. Since I ate $12 = 3 \times 4$ ice cream bars, I ate 4 yogurt bars.
   A) 4   B) 9   C) 15   D) 36

   9. A

10. $31 - 10 = 21$; so subtract 10 three times.
    A) 10   B) 11   C) 21   D) 30

    10. D

11. Add $10 + 20 + 30 + 40 = 100$.
    A) 10   B) 40   C) 100   D) 110

    11. C

12. $3 \times 35 = 105$; so $106 \div 3 = 35$ with remainder $106 - 105 = 1$.
    A) 0   B) 1   C) 2   D) 3

    12. B

Go on to the next page ⇒ **4**

		Answer Column
13.	$5\times4\times3\times2$ is divisible by 4. Only choice A has no factor of 4. A) $5\times3\times3\times2$ B) $4\times2\times5\times3$ C) $5\times2\times4\times3$ D) $5\times4\times3\times2$	13. A
14.	List all multiples of 36 until you find one which is divisible by 8. The multiples of 36 are 36, 72, . . . , and 72 is divisible by 8. A) 4 B) 36 C) 72 D) 288	14. C
15.	Since my school has 5 sets of twins, my school has $5\times2 = 10$ student twins. A) 5 B) 10 C) 15 D) 20	15. B
16.	The 1st, 12th, and 20th letters of the alphabet, in that order, are A, L, and T. A) Alex Louis Thomas B) Amy Lara Sanchez C) Anna Maria Trunk D) Albert Kevin Upton	16. A
17.	The 25 numbers 2×1, 2×2, 2×3, . . . , 2×25 are all divisible by 2. A) 23 B) 24 C) 25 D) 26	17. C
18.	12 tens = 120 & 12 ones = 12. Sum = 132 = 1 hundred + 32 ones. A) 12 B) 21 C) 22 D) 32	18. D
19.	All sides of a square have the same length, so the sum of the lengths of 3 sides is $3+3+3 = 9$. A) 3 B) 6 C) 9 D) 12	19. C
20.	The chef used 8 red plates, 6 green plates, and 5 white plates, so $8 + 6 + 5 = 19$ of today's *Chef's Specials* were *not* served on blue plates. A) 4 B) 18 C) 19 D) 23	20. C
21.	8 nickels + 7 dimes = 7 nickels + 1 nickel + 7 dimes = 7 nickels + 75¢, and 75¢ = 3 quarters. A) 2 B) 3 C) 6 D) 8	21. B
22.	Even multiples of 5 are also multiples of 10: 10, 20, 30, and 40. A) two B) four C) eight D) twenty	22. B

Go on to the next page ⫸ **4**

87

23. A radius is 2, so a diameter is 4. The diameter is twice as long as a side of the square, so a side of the square is 2. A) 1 B) 2 C) 4 D) 8	23. B
24. The 1st century was our calendar's first 100 years. The year 2001 A.D. is in the 21st century. A) 19th B) 20th C) 21st D) 200th	24. C
25. $3\times57+2 = 173$, $3\times72+1 = 217$, $3\times121+1 = 364$, and $3\times140+0 = 420$, so remainders are 2, 1, 1, and 0. A) 173 B) 217 C) 364 D) 420	25. A
26. Add two numbers from: 1, 3, 5, 7, 9. The result is always even. A) 2 B) prime C) odd D) even	26. D
27. Sides of 12 3-sided $\triangle$s = 12×3 = 36 = sides of 9 4-sided rectangles. A) 4 B) 8 C) 9 D) 16	27. C
28. My coins are worth \$8. Since each of the four coin-types has the same value, we know that the value of each type is $\$8 \div 4 = \$2$. Since I have \$2 in nickels, I must have exactly 40 nickels. A) 200 B) 40 C) 20 D) 8	28. B
29. My dog walked 10 times as far as Frank's, and Frank's walked 10 times as far as Al's. If Frank's walked 10 km, then mine walked 100 km, and Al's walked 1 km. The difference between the distances is $(100-1)$ km. A) 9 B) 90 C) 99 D) 109	29. C
30. Four straight lines can cross in as many as six points. A fifth line can cross the four other lines in four more points, as shown, for a total of 10. A) 9 B) 10 C) 12 D) 20	30. B

The end of the contest ✍ **4**

Visit our web site at http://www.mathleague.com

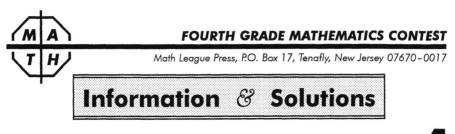

Information & Solutions

Spring, 2001

Contest Information

4

- **Solutions** Turn the page for detailed contest solutions (written in the question boxes) and letter answers (written in the *Answer Column* to the right of each question).

- **Scores** Please remember that *this is a contest, not a test*—and there is no "passing" or "failing" score. Few students score as high as 24 points (80% correct). Students with half that, 12 points, *deserve commendation!*

- **Answers & Rating Scale** Turn to page 142 for the letter answers to each question and the rating scale for this contest.

1. $20 \times 10 \times 2 \times 1 = 200 \times 2 \times 1 = 400 \times 1 = 400$.

   A) 212 B) 221 C) 400 D) 420

   1. C

2. 20 more than 300 is 320, and 1 more than 320 is 321.

   A) 32 B) 51 C) 312 D) 321

   2. D

3. $9+1 + 8+2 + 7+3 = 10+10+10 = 30 = 6+24$.

   A) 4 B) 24 C) 30 D) 36

   3. B

4. 10 quarters $= 10 \times 25¢ = 250¢ = 25 \times 10¢$.

   A) 5 B) 10 C) 15 D) 25

   4. D

5. 56 days $= (56 \div 7)$ weeks $= 8$ weeks.

   A) 6 B) 7 C) 8 D) 9

   5. C

6. There are 4 vowels (shown in bold) in "mathematics."

   A) 11 B) 7 C) 4 D) 3

   6. C

7. Twenty thousand + one hundred one $= 20\,000 + 101 = 20\,101$.

   A) 2101 B) 20 101 C) 21 101 D) 201 001

   7. B

8. 3 hrs $= 2$ hrs $+ 60$ mins $= 2$ hrs $+ 10$ mins $+ 50$ mins.

   A) 30 B) 50 C) 60 D) 90

   8. B

9. 10 less than 2001 is 1991, and 1991 is 10 more than 1981.

   A) 1981 B) 1991 C) 2001 D) 2011

   9. A

10. At the picnic, Sue swallowed 1 of every 6 seeds in her slice of watermelon. Sue must have swallowed $162 \div 6 = 27$ seeds.

    A) 27 B) 28 C) 52 D) 156

    10. A

11. $101 \times 10 \times 1 \times 0 \times 1 \times 10 \times 101 = 0 = 1010 \times 0$.

    A) 0 B) 1 C) 2 D) 3

    11. A

12. $21 \times 21 = 7 \times 3 \times 7 \times 3 = 7 \times 7 \times 3 \times 3 = 7 \times 7 \times 9$.

    A) 3 B) 7 C) 9 D) 21

    12. C

Go on to the next page ⟱ **4**

13. 1 dozen pairs = $1 \times 12 \times 2 = 24 = 8 \times 3$ or 8 triples.

A) 6   B) 8   C) 12   D) 36

14. A square has 4 sides, so the sum of the lengths of its sides equals the length of 1 side multiplied by 4.

A) 2   B) 3   C) 4   D) 8

15. 11 weeks + 11 days = (11×7) days + 11 days = 88 days = (88×24) hours = 2112 hours.

A) 22   B) 88   C) 121   D) 2112

16. 2 rocks are worth 5 stones, so 5×2 rocks are worth 5×5 stones.

A) 2   B) 4   C) 25   D) 50

17. $(32-24) + (32-24) + (32-24) + (32-24) = 8+8+8+8 = 4 \times 8$.

A) 4   B) 8   C) 24   D) 32

18. (# sides in a rectangle) − (# sides in a square) = 4 − 4 = 0 = (# of sides in a triangle) − 3.

A) 0   B) 1   C) 2   D) 3

19. $145+149 = 147+147$ & $146+148 = 147+147$: five 147s altogether.

A) 146   B) 147   C) 148   D) 149

20. Use guess and check. Increase the larger # and decrease the smaller # until the numbers differ by 31. Try 51 & 50, 52 & 49, . . . , 65 & 36, 66 & 35. I missed Dad 66 times.

A) 35   B) 49   C) 66   D) 70

21. Choice A, 66, is not divisible by 9.

A) 33 + 33   B) 6 + 6 + 6 + 6 + 6 + 6
C) 9 + 9 + 9 + 9 + 9   D) 18 + 18

22. The remainders are 4 and 9. Their sum is 4 + 9 = 13.

A) 4   B) 5   C) 9   D) 13

Go on to the next page ⏭ **4**

23. Try examples like 2×2 and 3×3. Choice A is correct.

A) 1, but not necessarily 2   B) 2, but not necessarily 1
C) 1 & 2, but not necessarily 3   D) 1, 2, & 3

23.

A

24. (sum of all digits in 2000)$\times$(sum of all digits in 2001) $= 2 \times 3 = 6$.

A) 0   B) 2   C) 3   D) 6

24.

D

25. A dragon grows 1 m each day or 7 m each week. A dinosaur grows 2 m each week. In 4 weeks, a dragon grows 28 m and a dinosaur grows 8 m.

A) 4   B) 8   C) 12   D) 20

25.

D

26. Each diameter is 2 radii, so 4 diameters = 8 radii. Since 8 radii = 128, one radius = $128 \div 8 = 16$.

A) 4   B) 8   C) 16   D) 32

26.

C

27. The rectangle, 3 squares wide$\times$7 squares long, can hold $3 \times 7 = 21$ squares. Or, area of rectangle $= 6 \times 14 = 84$, and $84 \div 4 = 21$ squares.

A) 14   B) 21   C) 28   D) 49

27.

B

28. For I and IV, use first letter of each word; for II, use first two letters of first word and first letter of other words; and for III, use first three letters of first word and first letter of second word. MATH could be an acronym for all four word groups.
   I. Multiply All Those Hundreds
   II. MArtians Take Hostages
   III. MATthew Hides
   IV. Minutes After The Hour

A) one   B) two   C) three   D) four

28.

D

29. Each of the 5 target stripes on my shirt is worth a different odd whole number less than 50. The greatest possible sum is $49+47+45+43+41 = 225$.

A) 220   B) 225   C) 235   D) 245

29.

B

30. $2001+(2000-1999)+(1998-1997)+ \ldots +(2-1) = 2001+(1000 \text{ ones})$.

A) 2001   B) 3001   C) 4001   D) 4002

30.

B

The end of the contest ✍ **4**

5th Grade Solutions

1996-1997 through 2000-2001

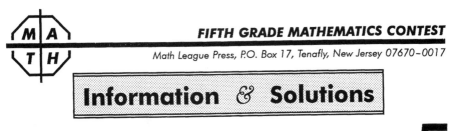
Information & Solutions

Spring, 1997

Contest Information

5

- **Solutions** Turn the page for detailed contest solutions (written in the question boxes) and letter answers (written in the *Answer Column* to the right of each question).

- **Scores** Please remember that *this is a contest, not a test*—and there is no "passing" or "failing" score. Few students score as high as 24 points (80% correct). Students with half that, 12 points, *deserve commendation!*

- **Answers & Rating Scale** Turn to page 143 for the letter answers to each question and the rating scale for this contest.

	Answer Column

1. $5+10+15=(4+1)+(9+1)+(14+1)=(4+9+14)+(1+1+1)=4+9+14+3.$

   A) 2   B) 3   C) 4   D) 5

   1. B

2. I found 3 blue, 4 red, and 2 white marbles, for a total of 9 marbles that aren't green. Those 9 marbles are 4 more than the 5 green marbles that I found.

   A) 4   B) 5   C) 9   D) 14

   2. A

3. 5 nickels + 5 dimes is worth $(5 \times 5¢) + (5 \times 10¢) = 25¢ + 50¢ = 75¢ = 3$ quarters.

   A) 2   B) 3   C) 4   D) 5

   3. B

4. If you divide by 7, all the 7's become 1's.

   A) 7   B) 8   C) 9   D) 10

   4. A

5. Any product involving one or more even numbers is even.

   A) even   B) odd   C) more than 10   D) prime

   5. A

6. 19 hundreds + 8 tens + 17 ones = 1900 + 80 + 17 = 1997.

   A) 1987   B) 1996   C) 1997   D) 2717

   6. C

7. If my art class starts at 2:45 P.M. and ends at 4:15 P.M., my art class lasts 15 minutes + 1 hour + 15 minutes = 90 minutes.

   A) 60   B) 75   C) 90   D) 120

   7. C

8. 500 cm + 50 m + 5 km = 5 m + 50 m + 5000 m = 5055 m

   A) 55   B) 550   C) 555   D) 5055

   8. D

9. If the number of marchers in the last parade was equal to the largest even number less than 2000, the number of marchers was 2000 − 2 = 1998.

   A) 1000   B) 1998   C) 1999   D) 2000

   9. B

10. In a week, there are $7 \times 24 = 168$ hrs.

    A) 7   B) 24   C) 168   D) 10 080

    10. C

11. A Pizza Heaven pizza costs $10.80 and is always cut into 8 equal slices. The cost per slice is $10.80 ÷ 8 = $1.35.

    A) $1.10   B) $1.20   C) $1.25   D) $1.35

    11. D

Go on to the next page ⏵ **5**

12. $6 \times (3 \times 5 \times 7) - 1(3 \times 5 \times 7) = (6-1) \times (3 \times 5 \times 7) = 5 \times (3 \times 5 \times 7)$

A) 7   B) 6   C) 5   D) 4

12.
C

---

13. This is multiplication by $2 \times 5 = 10$, so the product ends in a 0.

A) 0   B) 1   C) 2   D) 5

13.
A

---

14. 1 quarter+9 dimes+9 nickels+7 pennies=25¢+90¢+45¢+7¢=$1.67.

A) $1.67   B) $1.97   C) $2.52   D) $19.97

14.
A

---

15. The six 3-digit numbers that use *all* the digits
1, 2, & 3 are 123, 132, 213, 231, 312, & 321.

A) 3   B) 4   C) 5   D) 6

15.
D

---

16. The tennis ball was moving at 240 km/hr =
(240 km)/(60 min) = 4 km/min.

A) 2   B) 4   C) 6   D) 8

16.
B

---

17. The largest product appearing below is 75.

A) $2 \times 4 \times 9 = 72$ B) $2 \times 3 \times 12 = 72$
C) $3 \times 4 \times 6 = 72$ D) $3 \times 5 \times 5 = 75$

17.
D

---

18. One perimeter is twice the other, so if the larger square's side
is 6 cm, the smaller square's side is 3 cm and its area is 9 cm^2.

A) 9 cm^2 B) 18 cm^2 C) 36 cm^2 D) 72 cm^2

18.
A

---

19. Since there are $1 \times 4 = 4$ quarters in 1 dollar, there
are $10 \times 4 = 40$ quarters in 10 dollars.

A) 25   B) 40   C) 50   D) 250

19.
B

---

20. If the number of hats my aunt wears is 1 less
than the thousands' digit of 17 854, then the
number of hats she wears is 1 less than 7; it's 6.

A) 9   B) 8   C) 7   D) 6

20.
D

---

21. If Jo is the 2nd tallest of 126 fifth grade students,
she is taller than $126 - 2 = 124$ of these students.

A) 123   B) 124   C) 125   D) 126

21.
B

---

22. At a cost of $1 for the first minute and 75¢ for the other 4
minutes, a 5-minute call costs $1 \times \$1 + 4 \times 75¢ = \$1 + \$3 = \$4$.

A) $3.25   B) $3.75   C) $4   D) $5

22.
C

Go on to the next page ⏵ **5**

23. 176/3 has quotient 58, remainder 2; 173/5 has quotient 34, remainder 3. The other remainders: B is a 2, C is a 1, and D is a 0.

    A) $173 \div 5$ B) $174 \div 4$ C) $175 \div 3$ D) $176 \div 2$

    23.

    A

24. The unequal #s, each < 10, could be $0+1+3 = 4$, $1+2+3 = 6$, or $7+8+9 = 24$. A greater sum is *not* possible.

    A) 4   B) 6   C) 24   D) 25

    24.

    D

25. I sold half, you sold half the remainder. The unsold part was 40 books, so your "half" was 40 books. My "half" was twice yours. We began with 160 books.

    A) 10   B) 80   C) 120   D) 160

    25.

    D

26. The pogs I buy double in number each week. I bought 1 pog last week, 2 this week, 4 next week, then 8, then 16. The total is $1+2+4+8+16 = 31$.

    A) 31   B) 16   C) 15   D) 7

    26.

    A

27. Since the window (which is twice as high as it is wide) is 30 cm wide, the window must be 60 cm high. The sum is 90 cm.

    A) 45 cm   B) 60 cm   C) 90 cm   D) 120 cm

    27.

    C

28. Only 1 of 3 doesn't dance. The group of 120 is 3 groups of 40, and only 1 group of 40 does not dance.

    A) 30   B) 40   C) 60   D) 80

    28.

    B

29. The sum is 600. The 5-number average is 120, so the middle number, the third of the five numbers, is 120. The first is 2 less: it's 118.

    A) 596   B) 120
    C) 118   D) 116

    29.

    C

30. Jack and Jill each added 1000 numbers (Jack added the odds and Jill added the evens), and Jill got 1 more than Jack each time.

    A) 999   B) 1000   C) 1999   D) 2000

    30.

    B

The end of the contest ✍ **5**

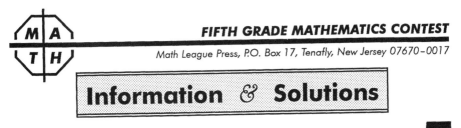
Information & Solutions

Spring, 1998

Contest Information

5

- **Solutions** Turn the page for detailed contest solutions (written in the question boxes) and letter answers (written in the *Answer Column* to the right of each question).

- **Scores** Please remember that *this is a contest, not a test*—and there is no "passing" or "failing" score. Few students score as high as 24 points (80% correct). Students with half that, 12 points, *deserve commendation!*

- **Answers & Rating Scale** Turn to page 144 for the letter answers to each question and the rating scale for this contest.

1. $5+10+15+20+25 = 1\times5 + 2\times5 + ... + 5\times5 = (1+2+3+4+5)\times5.$
   A) 2   B) 3   C) 4   D) 5

   1.
   D

2. One basket can hold a half-dozen eggs. Since 6 halves = 3 wholes, I'd need 6 baskets to hold 3 dozen eggs.
   A) 3   B) 4   C) 5   D) 6

   2.
   D

3. The largest angle in a right triangle is the right angle. Its measure is 90°.
   A) 45°   B) 60°   C) 90°   D) 180°

   3.
   C

4. To see which value is different, count the total number of 0's in each product.
   A) 40×50 B) 4×5000 C) 50×400 D) 40×500

   4.
   A

5. In division by 4, the last 2 digits determine the remainder; use $22\div4.$
   A) 3   B) 2   C) 1   D) 0

   5.
   B

6. One bag costs 75¢. Three such bags cost $3\times\$0.75 = \$2.25.$
   A) $0.25   B) $1.50   C) $2.25   D) $3.00

   6.
   C

7. Keep dividing each by 2 until you get an odd number.
   A) 2×15 B) $2\times2\times2\times2\times2\times1$ C) $2\times2\times9$ D) $2\times2\times2\times5$

   7.
   A

8. Since he will be 21 days old in 2 days, he is $21 - 2 = 19$ days old today.
   A) 18   B) 19   C) 20   D) 21

   8.
   B

9. Add five 20's to $10+11+12+13+14$ to get $30+31+32+33+34$. The answer is D.
   A) 5   B) 20   C) 3×20 D) 5×20

   9.
   D

10. $1000 + 900 + 80 + 18 = 1980 + 18 = 1998.$
    A) 1918   B) 1988   C) 1998   D) 19 818

    10.
    C

11. Since $162 \div 18 = 9$, we want a quotient of 10. That's D.
    A) $128 \div 16$ B) $120 \div 15$ C) $132 \div 12$ D) $110 \div 11$

    11.
    D

12. (# of angles in a square) + (# of angles in a $\triangle$) = $4 + 3 = 7.$
    A) 7   B) 9   C) 10   D) 12

    12.
    A

Go on to the next page ⇒ **5**

13. The largest product of a 3-digit number and a 2-digit number is $999 \times 99 = 98\,901$. That product has 5 digits.

    A) 3   B) 4   C) 5   D) 6

    13. C

14. $200 \div 7 > 28$, so the largest such multiple is $28 \times 7 = 196$.

    A) 187   B) 189   C) 196   D) 197

    14. C

15. Dancing pencils cost 74¢ each for the first dozen and 69¢ each for the rest, so 27 such pencils cost $12 \times \$0.74 + 15 \times \$0.69 = \$8.88 + \$10.35 = \$19.23$.

    A) $18.63   B) $19.23   C) $19.38   D) $19.98

    15. B

16. There are 60 seconds in each minute, so the quotient of the two quantities is 60.

    A) 60   B) 420   C) 3600   D) 7200

    16. A

17. 60 cm/hr = 1 cm/min = 100 cm/100 mins = 1 m/100 mins.

    A) 90 cm   B) 1 m   C) 110 cm   D) 120 cm

    17. B

18. The average of any 9 numbers is their sum divided by 9.

    A) $9 + 1998$ B) 9×1998 C) $1998 \div 9$ D) $9 \div 1998$

    18. C

19. Each minute, the second hand goes once around the face of Mom's clock. When Mom jogs for 5 minutes, the second hand goes around her clock 5 times.

    A) 5   B) 30   C) 60   D) 300

    19. A

20. A polygon has as many angles as it has sides.

    A) 3   B) 5   C) 8   D) 10

    20. B

21. The least common multiple of 6 and 18 is 18. The greatest common factor of 6 and 18 is 6. Finally, $6 \times 18 = 108$.

    A) 6   B) 18   C) 54   D) 108

    21. D

22. The area of a 2 by 4 rectangle is eight, so it can be divided into eight 1 by 1 squares.

    A) 2   B) 4   C) 6   D) 8

    22. D

Go on to the next page ⫸ **5**

23. Since 7 pencils cost as much as 3 pens, 6 times as many pencils cost as much as $6 \times 3 = 18$ pens.

A) 6   B) 18   C) 21   D) 98

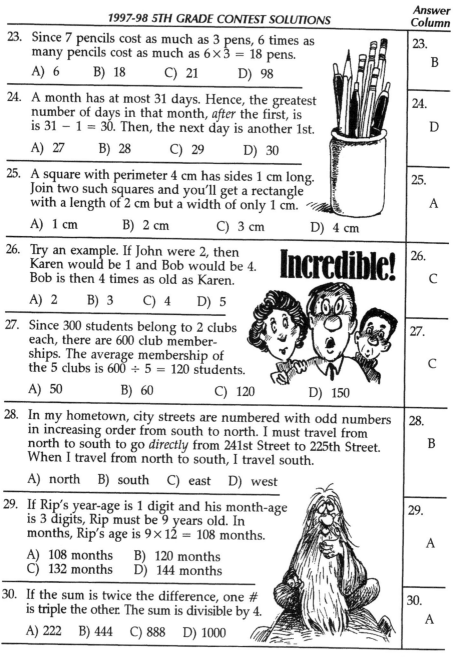

23.

B

24. A month has at most 31 days. Hence, the greatest number of days in that month, *after* the first, is is $31 - 1 = 30$. Then, the next day is another 1st.

A) 27   B) 28   C) 29   D) 30

24.

D

25. A square with perimeter 4 cm has sides 1 cm long. Join two such squares and you'll get a rectangle with a length of 2 cm but a width of only 1 cm.

A) 1 cm   B) 2 cm   C) 3 cm   D) 4 cm

25.

A

26. Try an example. If John were 2, then Karen would be 1 and Bob would be 4. Bob is then 4 times as old as Karen.

Incredible!

A) 2   B) 3   C) 4   D) 5

26.

C

27. Since 300 students belong to 2 clubs each, there are 600 club member-ships. The average membership of the 5 clubs is $600 \div 5 = 120$ students.

A) 50   B) 60   C) 120   D) 150

27.

C

28. In my hometown, city streets are numbered with odd numbers in increasing order from south to north. I must travel from north to south to go *directly* from 241st Street to 225th Street. When I travel from north to south, I travel south.

A) north   B) south   C) east   D) west

28.

B

29. If Rip's year-age is 1 digit and his month-age is 3 digits, Rip must be 9 years old. In months, Rip's age is $9 \times 12 = 108$ months.

A) 108 months   B) 120 months
C) 132 months   D) 144 months

29.

A

30. If the sum is twice the difference, one # is triple the other. The sum is divisible by 4.

A) 222   B) 444   C) 888   D) 1000

30.

A

The end of the contest ✍ **5**

Information & Solutions

Spring, 1999

Contest Information

5

- **Solutions** Turn the page for detailed contest solutions (written in the question boxes) and letter answers (written in the *Answer Column* to the right of each question).

- **Scores** Please remember that *this is a contest, not a test*—and there is no "passing" or "failing" score. Few students score as high as 24 points (80% correct). Students with half that, 12 points, *deserve commendation!*

- **Answers & Rating Scale** Turn to page 145 for the letter answers to each question and the rating scale for this contest.

1. Four groups of (1999 + 2001) = 4 × (4000) = 16 000.

   A) 4000   B) 8000   C) 12 000   D) 16 000

   1. D

2. $(1 \times 10) + (2 \times 100) + (3 \times 1) = 10 + 200 + 3 = 213$.

   A) 123   B) 213   C) 231   D) 312

   2. B

3. Since $27 \div 3 = 9$, choice C is correct.

   A) $25 \div 3$ B) $26 \div 3$ C) $27 \div 3$ D) $28 \div 3$

   3. C

4. At *Soup From Doc*, it costs me a quarter for each tablespoon of soup I buy. Since 7 tablespoons is $7 - 4 = 3$ more than 4 tablespoons, the 3 additional tablespoons cost me 3 more quarters.

   A) 3   B) 7   C) 75   D) 175

   4. A

5. Dividing out the common factor of 24, $(2 \times 4 \times 6 \times 8) \div (1 \times 2 \times 3 \times 4) = 2 \times 8 = 16$.

   A) 2   B) 4   C) 8   D) 16

   5. D

6. Use ones' digits: 6×7×8 ends in a 6.

   A) 8   B) 6   C) 4   D) 1

   6. B

7. Today is Tues. In 7 days it's Tues., so in 8 days it's Wed.

   A) Monday   B) Tuesday   C) Wednesday   D) Thursday

   7. C

8. Ten-thousands' digit of 654 321 + tens' digit of 654 321 = 5+2 = 7.

   A) 8   B) 7   C) 6   D) 5

   8. B

9. A year's supply of Frisbees cost me $48.72. Average monthly cost is $48.72 ÷ 12 = $4.06.

   A) $4.06   B) $4.60   C) $4.66   D) $6.00

   9. A

10. $1111 \div 22 = 50R11$, so quotient × remainder = $50 \times 11 = 550$.

    A) 25   B) 250   C) 550   D) 1100

    10. C

11. The total number of sides for 1 triangle + 1 rectangle + 1 hexagon = 3+4+6 = 13.

    A) 3   B) 11   C) 12   D) 13

    11. D

Go on to the next page ▐▶ **5**

12. $10\,000 \div 9 = (9999+1) \div 9 = 1111$ with remainder 1.

   A) 1   B) 3   C) 5   D) 7

13. The first 6 months take $31+28$ (or 29)$+31+$ $30+31+30 = 181$ (or 182). Day 200 is in July.

   A) June   B) July
   C) August   D) September

14. 753 is 357 more than $753-357 = 396$.

   A) 396   B) 404   C) 406   D) 1110

15. $1300 + 130 + 13 = 1443$.

   A) 333   B) 1333   C) 1433   D) 1443

16. Avg of 1 & 5 is 3. Avg of 1 and 7 is 4. Both are whole numbers.

   A) odd   B) even   C) prime   D) whole

17. A's value is $2.25, B's is $1.10, C's is $1.25, and D's is $1.00.

   A) 45 nickels   B) 11 dimes   C) 5 quarters   D) 1 dollar

18. I spent 1/3 of $120. I spent $40. I have $120-$40 = $80 left.

   A) $40   B) $60   C) $80   D) $90

19. The difference between Grandpa's height and mine is 123 cm. Grandpa's height is 202 cm. My height is $202-123 = 79$ cm.

   A) 77   B) 79   C) 87   D) 89

20. I colored every Sun, Tues, Thurs, & Sat. Each week, 3 days stay uncolored. In 7 weeks, 21 days stay uncolored.

   A) 49   B) 35   C) 28   D) 21

21. Since 23 is 17 more than twice the 2nd number, twice the 2nd number is 6, and the 2nd number is 3. First + 2nd = $23+3 = 26$.

   A) 26   B) 40   C) 73   D) 96

22. Only 6 has 4 different divisors (1, 2, 3, and 6). The others have only 3 divisors: 4 has 1, 2, 4; 9 has 1, 3, 9; 25 has 1, 5, 25.

   A) 4   B) 6   C) 9   D) 25

Go on to the next page ⮕ **5**

23. $(1 + 2 + 3 + 4) + (5 \times 2) + 10 = 10 + 10 + 10$. That's 3 tens.

   A) 0   B) 1   C) 3   D) 30

23.

C

24. A single costs 50¢. A double costs 75¢. Now, $5 = 6 \times 75¢ + 50¢$, so $5 can buy at most 6 doubles and 1 single = 13 blasts.

   A) 10   B) 12   C) 13   D) 15

24.

C

25. $20 \times 30 \times 40 = (2 \times 10) \times (6 \times 5) \times (4 \times 10) = 2 \times 6 \times 4 \times 500$.

   A) 10   B) 50   C) 100   D) 500

25.

D

26. To find the missing number in $5:7 = \underline{\ ?\ }:28$, multiply the first ratio by 4. This gives you $5:7 = (4 \times 5):(4 \times 7) = 20:28 = \underline{\ ?\ }:28$, so the answer is 20.

   A) 15   B) 20   C) 23   D) 25

26.

B

27. $(14 \times 11) + (13 \times 11) + (12 \times 11) + (11 \times 11) = (14 + 13 + 12 + 11) \times 11 = 50 \times 11$.

   A) 40   B) 50   C) 60   D) 61

27.

B

28. The 20 guests at Tropical Island got different whole numbers from 1 to 20. Whenever two guests added up to 21, they could share a hammock. Pair 1 & 20, 2 & 19, 3 & 18, . . . , 10 & 11—that's 10 pairs in all.

   A) 10   B) 11   C) 20   D) 21

28.

A

29. To avg 95, she needs $5 \times 95 = 475$ total points. Her first 3 grades total 281, so she needs 194 more points on her next 2 tests, for an average of $194 \div 2 = 97$.

   A) 97   B) 98   C) 99   D) 100

29.

A

30. The primes are 53, 59, 61, 67, 83, 89, 97. The others are composite.

   A) 50 and 60   B) 60 and 70   C) 80 and 90   D) 90 and 100

30.

D

The end of the contest ☝ **5**

Information & Solutions

Spring, 2000

Contest Information

5

- **Solutions** Turn the page for detailed contest solutions (written in the question boxes) and letter answers (written in the *Answer Column* to the right of each question).

- **Scores** Please remember that *this is a contest, not a test*—and there is no "passing" or "failing" score. Few students score as high as 24 points (80% correct). Students with half that, 12 points, *deserve commendation!*

- **Answers & Rating Scale** Turn to page 146 for the letter answers to each question and the rating scale for this contest.

1. $(2+10)+(4+10)+(6+10)+(8+10) = 2+4+6+8 + 40.$
   A) 10   B) 20   C) 30   D) 40

   1.
   D

2. 17 days before Tues. = 2 weeks & 3 days before Tues. = 3 days before Tues. = Sat.
   A) Thur.   B) Fri.   C) Sat.   D) Sun.

   2.
   C

3. $100 \div 5 = 20 = 4 \times 5.$
   A) 2   B) 4   C) 10   D) 20

   3.
   B

4. Since the smallest whole number is 0, the product equals 0.
   A) 0   B) 15   C) 120   D) 121

   4.
   A

5. The ones' digit is 0. That's 8 less than the hundreds' digit of 8.
   A) 0   B) 1   C) 7   D) 8

   5.
   D

6. Use an example: $20-5 = 15; 20+5 = 25.$ The sum is always odd.
   A) a one-digit   B) a prime   C) an even   D) an odd

   6.
   D

7. If 1 quarter = 5 nickels, then 5 quarters = (5×5) nickels.
   A) 15   B) 20   C) 25   D) 50

   7.
   C

8. Gil the Fish weighs twice as much as Bill the Fisherman. If Gil weighs 150 kg, then Bill weighs $(150 \div 2)$ kg.
   A) 75   B) 150   C) 225   D) 300

   8.
   A

9. $2+22+222 = 246 = 2 \times 123.$
   A) $1 + 11 + 111 = 123$ B) $1 + 10 + 110 = 121$
   C) $2 + 12 + 112 = 126$ D) $1 + 12 + 24 = 37$

   9.
   A

10. (# of digits in 10 000 000)$\div$(# of digits in 1000) = $8 \div 4 = 2.$
    A) 2   B) 8/3   C) 4   D) 10 000

    10.
    A

11. (The number of sides in a triangle) + (the number of sides in a pentagon) = $3 + 5 = 8$ = the number of sides in an octagon.
    A) a square   B) a rhombus   C) a hexagon   D) an octagon

    11.
    D

12. My secret number uses the digits 1, 2, and 3 once each. The only 6 possibilities are: 123, 132, 213, 231, 312, and 321.
    A) 3   B) 4   C) 5   D) 6

    12.
    D

Go on to the next page ⏭ **5**

1999-2000 5TH GRADE CONTEST SOLUTIONS

13. The difference 300−100 = 200, and 500 more than 200 is 700.

    A) 200   B) 300   C) 600   D) 700

14. 120 minutes − **75** minutes = 45 minutes.

    A) 30   B) 75   C) 85   D) 155

15. Kyle cried 3 crocodile tears each day.
    Kyle cried 3×7 = 21 tears last week.

    A) 7   B) 10   C) 15   D) 21

16. Add 555 555 555 555 555 to itself. The sum
    is 1 111 111 111 111 110, so there's one 0.

    A) 1   B) 5   C) 15   D) 29

17. In 1000÷7, the quotient is 142. Largest multiple is 7×142 = 994.

    A) 3   B) 4   C) 7   D) 9

18. A is divisible by 6. D is 7 more than A, and 7 = 6+1.

    A) 612 481 230 B) 612 481 239 C) 612 481 238 D) 612 481 237

19. No matter how many 4s are averaged, the average remains 4.

    A) twos   B) fours   C) sevens   D) eights

20. The width of the square is (36÷4) cm or 9 cm. The width of the
    rectangle is (36÷18) cm or 2 cm. Their widths differ by 7 cm.

    A) 4 cm   B) 7 cm   C) 32 cm   D) 34 cm

21. Add the 20 kids who are 10 or
    younger and the 8 kids who
    are older than 10 to get a sum
    of 28 kids. The kids older than
    11 are also older than 10, so
    they were already counted.

    A) 22   B) 26   C) 28   D) 34

22. Since 15 = 1×15 = 3×5,
    the 4 factors of 15 are 1, 3, 5, & 15.

    A) 11 = 1×11 B) 13 = 1×13 C) 15   D) 17 = 1×17

23. (40−39)+(38−37)+...+(4−3)+(2−1) = 1+1+ ... +1+1 = 20.

    A) 1   B) 20   C) 21   D) 40

Go on to the next page ⫸ **5**

109

	Answer Column
24. All sides are 5, so $AB + BC = 5 + 5 = AD + AB$. A) $AD + AC$ B) $BC + BD$ C) $AC + BC$ D) $AD + AB$	24. D
25. A rectangle could have side-lengths 8 cm, 8 cm, 16 cm, & 16 cm. A) rectangle B) square C) triangle D) circle	25. A
26. Together, the consonants and the vowels are the 26 letters of the alphabet. Each of the numbers from 1 to 26 will be assigned to a letter. The sum of these numbers is $1+2+3+ \ldots +24+25+26 = (1+26)+ (2+25)+ \ldots +(13+14) = 27+27+ \ldots +27 = 27 \times 13 = 351$. A) 300 B) 326 C) 330 D) 351	26. D
27. My 7 coins must be 2 pennies, 2 nickels, 2 dimes, and 1 quarter. A) penny B) nickel C) dime D) quarter	27. D
28. Try an example. The difference $19\,992\,000 - 19\,991\,999 =$ the difference $19\,992\,001 - 19\,992\,000$, and the average of $19\,992\,001$ and $19\,991\,999$ is $19\,992\,000$. The average is *always* $19\,992\,000$. A) $6\,664\,000$ B) $9\,996\,000$ C) $19\,992\,000$ D) $39\,984\,000$	28. C
29. Since the 7 ice sculptures were carved at the rate of one per hour, the first sculpture was finished at the end of the first hour, and the seventh was completed 6 hours later, at the end of the seventh hour. The first sculpture melted $6 \times 10 = 60$ cm during these 6 hours. A) 50 B) 60 C) 70 D) 80	29. B
30. The pattern around the circle was boy, boy, girl, boy, girl, boy, girl, boy, girl (then back to boy, boy, . . .). No girl stood next to another girl. A) 0 B) 1 C) 2 D) 3	30. A

The end of the contest ☝ **5**

Information & Solutions

Spring, 2001

Contest Information

5

- **Solutions** Turn the page for detailed contest solutions (written in the question boxes) and letter answers (written in the *Answer Column* to the right of each question).

- **Scores** Please remember that *this is a contest, not a test*—and there is no "passing" or "failing" score. Few students score as high as 24 points (80% correct). Students with half that, 12 points, *deserve commendation!*

- **Answers & Rating Scale** Turn to page 147 for the letter answers to each question and the rating scale for this contest.

1. 30 + 40 + 50 = 30+50 + 40 = 80 + 40.

    A) 50   B) 40   C) 30   D) 20

    1.
    B

2. 100 more than 50 is 150, and 50 less than 150 is 100.

    A) 25   B) 50   C) 75   D) 100

    2.
    D

3. If 2 worms crawl south every hour, then 6×2 = 12 worms crawl south in 6 hours.

    A) 3   B) 6   C) 8   D) 12

    3.
    D

4. 63 = 9×7 = (6+3)×7.

    A) 7   B) 8   C) 9   D) 10

    4.
    A

5. If every chocolate chip cookie contains 20 chocolate chips, then 100 chocolate chip cookies contain 100×20 = 2000 chocolate chips.

    A) 5   B) 120   C) 200   D) 2000

    5.
    D

6. 111−101 = 10, 111−109 = 2, 119−111 = 8, and 121−111 = 10.

    A) 101   B) 109   C) 119   D) 121

    6.
    B

7. An express train travels twice as fast as a local train on the same route. The local train travels the route in 6 hrs. The express travels twice as fast, so it takes half as long, 3 hours, to travel the route.

    A) 3   B) 4   C) 8   D) 12

    7.
    A

8. June 8 through June 30 is the same as June 1 through June 23. It's 23 days, & $10×23 = $230.

    A) $210   B) $220   C) $230   D) $300

    8.
    C

9. 10×20×30 = 1×2×3×10×10×10.

    A) 6   B) 10   C) 100   D) 1000

    9.
    D

10. A polygon has sides. A circle has no sides.

    A) circle   B) rectangle   C) square   D) triangle

    10.
    A

11. 11 tens + 11 ones = 11×10 + 11×1 = 110 + 11 = 121.

    A) 110   B) 111   C) 121   D) 122

    11.
    C

Go on to the next page ⫸ **5**

12. $5¢ + 50¢ + \$500 = \$0.05 + \$0.50 + \$500.00 = \$500.55.$ A) \$5.55 B) \$50.55 C) \$500.55 D) \$555.00	12. C
13. At 2400 words per hr., I average $(2400÷60) = 40$ words per min. A) 24 B) 40 C) 60 D) 120	13. B
14. Paul buys a new football every 3 years. Paul bought his 1st football when he was 8, his 2nd at 11, 3rd at 14, 4th at 17, and 5th at 20. A) 11 B) 19 C) 20 D) 23	14. C
15. The top and bottom are parallel and so are the other 2 sides. A) none B) 1 C) 2 D) 4	15. C
16. The average of 3, 5, 7, and 9 is $(3+5+7+9)÷4 = 24÷4 = 6.$ A) 5 B) 6 C) 7 D) 8	16. B
17. Since $3×6 = 18$ and $18×2 = 36$, choice C is correct. A) 18 B) 26 C) 36 D) 66	17. C
18. 20 dimes + 20 nickels $= 20×10¢ + 20×5¢ = 300¢ = 12$ quarters. A) 10 B) 12 C) 20 D) 300	18. B
19. 10 hours after 10 A.M. is 8 P.M., which is 10 hours before 6 A.M. A) 6 A.M. B) 4 A.M. C) 10 P.M. D) 8 P.M.	19. A
20. The last 5 digits of the product $110×120×130×140×150$ are all 0s. A) 0 B) 1 C) 5 D) 6	20. A
21. 12 hip-hops $= (6×2)$ hip-hops $= (6×3)$ hops $= (6×1)$ hips $= 6$ hips. A) 4 B) 6 C) 9 D) 18	21. B
22. Numbers like 17, 26, 35, 44, . . . will all leave a remainder of 8 when divided by 9. A) 0 B) 1 C) 7 D) 8	22. D

Go on to the next page ⏵ **5**

23.	Five years ago, Ike was 10, so Tina is now 10 + 10 = 20. In 2 years, Tina will be 22. A) 27　　B) 25　　C) 22　　D) 20	23. C
24.	7 days = (24 × 7) hrs = 168 hrs. A) 24　B) 168　C) 240　D) 336	24. B
25.	One possible way is to cut the paper in half, then cut one of these two pieces in half on a line that cuts both longer sides in half. A) 1　　　B) 2　　　C) 4　　　D) 8	25. B
26.	The remainders are A) 0, B) 1, C) 0, D) 2. Both 0 and 2 are even. A) 156 ÷ 12　B) 259 ÷ 3　C) 355 ÷ 5　D) 455 ÷ 3	26. B
27.	Drop the 0s: 1 × 2 × 3 × 4 = 24, a 2-digit number. A) 1　　　B) 2　　　C) 3　　　D) 4	27. B
28.	Together, Ann, Bob, Carl, & Dee weigh (180 + 210) kg = 390 kg. Since Ann & Carl weigh 220 kg, Bob & Dee weigh (390 − 220) kg = 170 kg. A) 170 kg　　B) 180 kg C) 190 kg　　D) 200 kg	28. A
29.	Make a list: 111, 212, 221, 313, 331, 414, 422, 441, 515, 551, 616, 623, 632, 661, 717, 771, 818, 824, 842, 881, 919, 933, and 991. A) 18　　　B) 19　　　C) 21　　　D) 23	29. D
30.	Putting 30 squares in a row creates a rectangle with length 4 × 30 = 120 and width 4. Its perimeter is 2 × (120 + 4) = 2 × 124 = 248. A) 88　　　B) 136　　　C) 248　　　D) 480	30. C

The end of the contest 🖎 **5**

114

6th Grade Solutions

• •

1996-1997 through 2000-2001

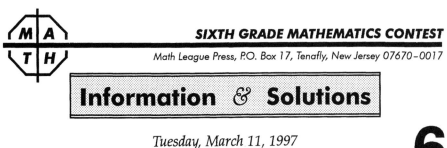
Information & Solutions

Tuesday, March 11, 1997

Contest Information

- **Solutions** Turn the page for detailed contest solutions (written in the question boxes) and letter answers (written in the *Answers* column to the right of each question).

- **Scores** Please remember that *this is a contest, not a test*—and there is no "passing" or "failing" score. Few students score as high as 30 points (75% correct). Students with half that, 15 points, *deserve commendation!*

- **Answers & Rating Scale** Turn to page 148 for the letter answers to each question and the rating scale for this contest.

1. As an example, days 1, 8, 15, 22, & 29 could be Sundays.
   A) 2 apples B) 3 apples C) 4 apples D) 5 apples

   1. D

2. $(1+19)\div2$ OR $(1+3+5+7+9+11+13+15+17+19)\div10$.
   A) 8   B) 9   C) 10   D) 11

   2. C

3. If the greatest common factor of two numbers is 15, their difference is also a multiple of 15.
   A) 60, 75 B) 75, 90 C) 90, 105 D) 105, 125

   3. D

4. $A = 32\div2$, $B = (32\times2)\div(2\times2)$, and $D = (32\times6)\div(2\times6)$ are equal.
   A) $32\div2 = 16$ B) $64\div4 = 16$ C) $128\div6 \neq 16$ D) $192\div12 = 16$

   4. C

5. 9×7 is common. Compare the rest: $A=1\times8$, $B=1\times9$, $C=0$, & $D=1\times2$.
   A) 1897   B) 1997   C) 2097   D) 2197

   5. B

6. Any number (except 0) divided by itself is equal to 1.
   A) 0   B) 1   C) 2   D) 3

   6. B

7. Multiply the ones' digits to get $5\times2 = 10$.
   A) 1   B) 5   C) 9   D) 0

   7. D

8. Every multiple of 6 is divisible by 3. If you add 1, the remainder is 1.
   A) 0   B) 1   C) 2   D) 3

   8. B

9. When you add an even number of odd numbers, the sum is even.
   A) even   B) odd   C) prime   D) less than 90

   9. A

10. 1 million $= 10^6$. There is no 10^6, so the millions' digit is a 0.
    A) 0   B) 1   C) 2   D) 3

    10. A

11. The least positive difference between unequal primes is $3-2 = 1$.
    A) 1   B) 2   C) 3   D) 4

    11. A

12. For half of the first 30 days in May, for 15 days, the date is an odd number. Since May 31 is an odd number date, Bobby was scared $15 + 1 = 16$ times.
    A) 15   B) 16   C) 17   D) 18

    12. B

13. 100010001000 is divisible by 2, 3, and 4.
    A) 3   B) 6   C) 9   D) 12

    13. C

14. Add 10 pennies to each of the choices. Only one will become a multiple of 25¢. That one is choice C.
    A) $1.35   B) $1.60   C) $1.65   D) $2.10

    14. C

15. If I need $105, I must withdraw the smallest multiple of $20 that's bigger than $105. That multiple is $120.
    A) $100   B) $120   C) $125   D) $130

    15. B

Go on to the next page Ⅲ➡ **6**

16.	I have $5. I spend 20% ($1) at the grocery. If I spend 50% of the $4 remaining ($2), I'll have $5 − $1 − $2 = $2 left. A) $1.50 B) $2 C) $3 D) $3.50	16. B
17.	Factor out all the 2's and rewrite as $1 \times 2 \times 3 \times 2^2 \times 5 \times (2 \times 3) \times 7 \times 2^3 \times 9 \times (2 \times 5)$. Remove 2's to get $1 \times 3 \times 5 \times 3 \times 7 \times 9 \times 5 = 14\,175$. (That's a whole lot of weeds!) A) 9 B) 567 C) 945 D) 14 175	17. D
18.	Primes: $2, 3, 5, 7, 11, 13, 17, 19, 23, 29, 31, 37$. A) 37 B) 31 C) 29 D) 23	18. A
19.	$3^3 \div 3 + 3^4 \div 3^2 + 3^5 \div 3^3 = 3^2 + 3^2 + 3^2 = 3 \times 3^2$. A) 1 B) 3 C) 3^2 D) 3^3	19. B
20.	In a day, the # of seconds is 60 times the # of minutes. A) 24 B) 60 C) 120 D) 3600	20. B
21.	The sum of all three angles is 180°, so each angle < 180°. A) 1° B) 120° C) 150° D) 180°	21. D
22.	The rectangle's perimeter is 20, so each side of the square is 5. A) 9 B) 20 C) 25 D) 81	22. C
23.	Since $8 \times 3 \times 4 = 96$, the operation is ×. A) + B) − C) ÷ D) ×	23. D
24.	Since diameter = 2 × radius, diameter : radius = 2:1. A) 1:2 B) 1:1 C) 2:1 D) 4:1	24. C
25.	The middle is 9.5, the avg. The 5 largest are 10, 11, 12, 13, and 14. A) 19 B) 15 C) 14 D) 10	25. C
26.	Start at the end and go backwards. If we each had $1 and Pat returned my 10¢, then Pat started with 90¢ and I started with $1.10. A) 5 B) 10 C) 15 D) 20	26. D
27.	The first 3 digits of Pizza Boat's phone number have the same sum as its last 4 digits. Only in choice A are all 7 digits different. A) 819-4536 B) 319-4503 C) 915-6054 D) 879-6574	27. A
28.	Since $10^{20} \div 10^{15} = 10^5$, the answer is C. A) 15 B) 4 C) 10^{15} D) 10^4	28. C
29.	$1997 \div 60 = 33.28\overline{3}$, and $33 \times 60 = 1980$. Finally, $1997 - 1980 = 17$. A) 17 B) 5 C) $0.28\overline{3}$ D) $0.208\overline{3}$	29. A

Go on to the next page ⟶ **6**

119

30. It's as though *every* math test is 12 points higher. The sum of the math tests would then be $6 \times 12 = 72$ points higher.
    A) 12   B) 24   C) 36   D) 72

30.
D

31. Since S is the 19th letter of the alphabet, and $18 \times 3 = 54$, the first 54 words began with one of the first 18 letters, not with S.
    A) 51st   B) 53rd
    C) 54th   D) 56th

31.
D

32. The sum of the 2 shortest sides is more than the longest side. The only possibilities are _?_,1,4 or _?_,2,3; so 1 isn't possible.
    A) 1 cm   B) 2 cm   C) 3 cm   D) 4 cm

32.
A

33. The area of each small square is 1, so the area of the unshaded rectangle is 2; and the area of each shaded triangle is half that, 1. Finally, $9 - 2 - 1 - 1 = 5$.
    A) 4   B) 4.5   C) 5   D) 5.5

33.
C

34. The least such number is the ten-digit number 1 999 999 999.
    A) 11   B) 10   C) 9   D) 8

34.
B

35. Subtract multiples of $7 from the choices until you get a multiple of $6. Here, $32 - $14 = $18, so $2 \times $7 + 3 \times $6 = $14 + $18 = $32.
    A) $22   B) $23   C) $29   D) $32

35.
D

36. 600% of 1 hr = (600/24)%, or 25%, of 1 day.
    A) 12   B) 24   C) 25   D) 50

36.
C

37. $\frac{1}{2} \times \frac{2}{3} \times \frac{3}{4} \times \ldots \times \frac{19}{20} = \frac{1}{20}$.
    A) $\frac{1}{20}$ B) $\frac{1}{10}$ C) $\frac{9}{10}$ D) $\frac{19}{20}$

37.
A

38. The smallest of the 50 is 1997, so the largest is 2046, and the average is 2021.5. Of the 25, the average number (which is the middle, or 13th, number) is $2 \times 2021.5 = 4043$. The smallest is just 12 less, 4031.
    A) 4032   B) 4031   C) 3994   D) 3993

38.
B

39. Group the eighty 2's into twenty groups of 2^4.
    A) 5th   B) 10th   C) 20th   D) 320th

39.
C

40. My clock advances 65 minutes every 60 minutes, a ratio of 65:60, or 13:12. When my clock advances the 13 hours from 6 P.M. to 7 A.M., an accurate clock advances 12 hours, to 6 A.M.
    A) 5:55   B) 6   C) 6:55   D) 8:05

40.
B

The end of the contest 🖎 **6**

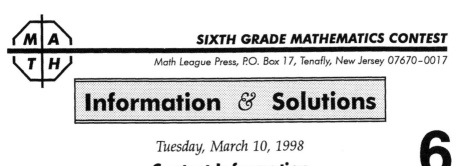

Information & Solutions

Tuesday, March 10, 1998

Contest Information

6

- **Solutions** Turn the page for detailed contest solutions (written in the question boxes) and letter answers (written in the *Answers* column to the right of each question).

- **Scores** Please remember that *this is a contest, not a test*—and there is no "passing" or "failing" score. Few students score as high as 30 points (75% correct). Students with half that, 15 points, *deserve commendation!*

- **Answers & Rating Scale** Turn to page 149 for the letter answers to each question and the rating scale for this contest.

1. If each car holds 5 clowns, then $60 \div 5 = 12$ cars hold 60 clowns.
   A) 5   B) 10   C) 12   D) 60

   1. C

2. $0 + 6 + 70 + 800 + 9000 = 9876$.

   A) 98765   B) 56789   C) 9876   D) 6789

   2. C

3. One week has $24 \times 7 = 168$ hours.
   A) 7   B) 24   C) 140   D) 168

   3. D

4. Moving the 8, $1+9+9+8 = 1+8+9+9$.
   A) $2+9+9+8$ B) $1+9+9+9$ C) $1+8+9+9$ D) $1+8+9+8$

   4. C

5. If you subtract the least positive odd number from the least positive even number, the result will be $2 - 1 = 1$.
   A) 0   B) 1   C) 2   D) 3

   5. B

6. The gcf of 40 and 80 is 40, and the gcf of 80 and 120 is also 40.
   A) 120   B) 140   C) 160   D) 400

   6. A

7. If 130 of 300 envelopes contain letters, then $300 - 130 = 170$ envelopes do *not* contain letters.

   We Get Letters..

   A) 130   B) 170   C) 270   D) 430

   7. B

8. The digit sum is always 9, so all the numbers are multiples of 9.
   A) 6   B) 9   C) 12   D) 18

   8. B

9. $(3 \times 1998) - 1998 = (3-1) \times 1998$.
   A) 1×1998 B) 2×1998 C) 3×1998 D) 4×1998

   9. B

10. The ones' digit is 5, so 305 gets rounded up to 310.
    A) 300   B) 310   C) 350   D) 400

    10. B

11. A square has 4 right angles and $4 \times 90° = 360°$.
    A) 90°   B) 180°   C) 360°   D) 400°

    11. C

12. $61 + 39 = 100$ and $71 + 29 = 100$; so $(3 \times 100) \div (3 \times 100) = 1$.
    A) $1 \div 2$ B) $2 \div 3$ C) $3 \div 4$ D) $5 \div 5$

    12. D

13. Together, my three cats weigh 60 kg. If the least weight of any of my cats is 20 kg, then each of my three cats must weigh 20 kg.
    A) 20   B) 30   C) 39   D) 40

    13. A

14. $(11+22+33) \div (1+2+3) = 66 \div 6 = 11$.
    A) 1×11 B) 3×11 C) 6×11 D) 6

    14. A

15. Reducing, $8{:}10 = (2 \times 4){:}(2 \times 5) = 4{:}5$.
    A) 12:10   B) 10:12   C) 10:8   D) 4:5

    15. D

16. The primes > 0 and < 40 are 2,3,5,7,11,13,17,19,23,29,31,37.
    A) 12   B) 13   C) 14   D) 15

    16. A

Go on to the next page ▐▶ **6**

17. A machine dispenses gumballs always in the order green, blue, yellow, orange, red. My first gumball was blue. I bought a total of 8 gumballs. The order was BYORGBYO, so my 8th gumball was orange. A) green B) blue C) yellow D) orange	17. D
18. 3 new kids joined 10 boys and 8 girls. The number of boys increased by 20%. There were then 12 boys & 9 girls. The new ratio of boys to girls was 4:3. A) 4:3 B) 3:4 C) 3:2 D) 2:3	18. A
19. $2^2+2^2+2^2+2^2 = 4+4+4+4 = 16$, while choice D is $2\times16 = 32$. A) $2\times2\times2\times2$ B) 4×2^2 C) 2^4 D) 2×4^2	19. D
20. If Square B's perimeter = 8 cm, its side = 2 cm and its area = $2^2 = 4$. The area of Square A, twice that of square B, is $2\times4 = 8$. A) 64 cm^2 B) 16 cm^2 C) 8 cm^2 D) 4 cm^2	20. C
21. A competition is held once every 4 years, so the competition can be held in years 1, 5, and 9. That's 3 times in a decade. A) 2 B) 3 C) 4 D) 5	21. B
22. If a pole was at the start of a 240 m road and every 12 m thereafter, poles appeared at 0, 12, . . . , 240, for 21 poles in all. A) 10 B) 12 C) 20 D) 21	22. D
23. Sniffles sneezed 30 minutes. Each sneeze took 30 seconds, so Sniffles sneezed twice a minute, or 60 times all together. A) 1 B) 10 C) 30 D) 60	23. D
24. The handshakes are: AB, AC, AD, AE, BC, BD, BE, CD, CE, and DE, a total of 10 different possible handshakes. A) 4 B) 5 C) 10 D) 20	24. C
25. They're consecutive, so average the middle two: $(6+7)\div2 = 6.5$. A) 6.5 B) 6 C) 7.5 D) 7	25. A
26. $2\times2 = 4 = 2 + 2$. A) 1 B) 2 C) 3 D) 4	26. B
27. Ali and I ran 60 m. If I ran twice as fast, then I ran twice as far. Ali ran 20 m and I ran 40 m. A) 10 B) 20 C) 30 D) 40	27. D
28. The Movie Theater sells two tickets for the price of one every Monday. The price of one ticket is $7. If quadruplets go to the movies on Monday, they must pay $2\times\$7 = \$14$. A) \$7 B) \$14 C) \$21 D) \$28	28. B
29. 7 oranges yield 7×10 ml = 70 ml, and 1 pineapple yields 30 ml, so the mixture is (30 ml)/(30 ml + 70 ml) = 30% pineapple. A) 25% B) 30% C) 70% D) 75%	29. B

Go on to the next page ⫸ **6**

30. $50¢ \times 24 \times 14 = \$168$ for two weeks at Pete's; $\$12 \times 14 = \$168$ for two weeks at Paul's, and $\$75 \times 2 = \$150$ for two weeks at Patty's. A) \$84　　B) \$140　　C) \$150　　D) \$168	30. C
31. A side of a square and a radius of a circle are both 4. The circle's area is 16π and the square's area is 16. A) 20π　B) 32π　C) $16\pi + 4$　D) $16\pi + 16$	31. D
32. A stick 42 cm long is divided into 6 equal parts by cuts made at the 7, 14, 21, 28, and 35 cm marks. A) 5　　B) 6　　C) 7　　D) 8	32. A
33. $10^2 - 10 \times 10 + 10 = 100 - 100 + 10 = 10$. A) 0　　B) 10　　C) 100　　D) 110	33. B
34. Since $1000 \div 17 \approx 58.8$, round down to say that 58 positive integers < 1000 are multiples of 17. A) 1　　B) $1000 \div 17$　C) 58　　D) 59	34. C
35. If Al scored 10, Bo scored 8, and Cy scored 6, then the ratio of Al's score to the average of Bo's and Cy's scores was 10:7. A) 10:7　　B) 7:10　　C) 5:3　　D) 3:5	35. A
36. The arrow traveled 100 cm/sec = 1 m/sec. Since the arrow traveled 50 m, Pat landed in 50 seconds. A) 0.5　B) 2　C) 50　D) 100	36. C
37. There can't be two factors of 3; but 333 has $3^2 = 9$ as a factor. A) 111　　B) 333　　C) 555　　D) 777	37. B
38. Peaches sold for \$1/kg. If 15 peaches cost \$4, then 15 peaches weighed 4 kg, and an average peach weighed 4/15 kg. A) $\frac{4}{15}$　B) $\frac{15}{4}$　C) $\frac{1}{15}$　D) 1	38. A
39. From 200 to 299 is 100 numbers. Each of the other 100's has 2 as a tens' digit 10 times, for a total of $100 + (9 \times 10) = 190$ times. A) 90　B) 100　C) 190　D) 200	39. C
40. Sun + 30 days = Tues; Tues + 31 days = Fri; Fri + 31 days = Mon, so the first month could be June or November. A) June　　B) July　　C) September　D) October	40. A

The end of the contest ✍ **6**

Information & Solutions

Tuesday, March 9, 1999

Contest Information

6

- **Solutions** Turn the page for detailed contest solutions (written in the question boxes) and letter answers (written in the *Answers* column to the right of each question).

- **Scores** Please remember that *this is a contest, not a test*—and there is no "passing" or "failing" score. Few students score as high as 30 points (75% correct). Students with half that, 15 points, *deserve commendation!*

- **Answers & Rating Scale** Turn to page 150 for the letter answers to each question and the rating scale for this contest.

Copyright © 1999 by Mathematics Leagues Inc.

1. Months with $\geq 4 \times 7 + 1 = 29$ days can have 5 Fridays. A month needs $5 \times 7 + 1 = 36$ days to have 6.
    A) 2   B) 3   C) 4   D) 5

    1. D

2. $(50 \times 60 \times 70 \times 80) \div (5 \times 6 \times 7 \times 8) = 10\,000$.
    A) 10   B) 100   C) 1000   D) 10 000

    2. D

    TOP BANANA

3. By calculator, $10\,101 = 3 \times 7 \times 13 \times 37$.
    A) 17   B) 13   C) 7   D) 3

    3. A

4. Commutative property of multiplication: $99 \times 101 = 101 \times 99$.
    A) 98×102 B) 97×103 C) 100×100 D) 101×99

    4. D

5. $24 \times 60 + 9 \times 60 + 19$ mins $= 33 \times 60 + 19$ mins $= 1999$ mins.
    A) 29   B) 583   C) 1999   D) 86 959

    5. C

6. Count: November, December, January, February.
    A) January   B) February   C) June   D) October

    6. B

7. The tens' and hundreds' digits of 9876 are 7 and 8, and $7 \times 8 = 56$.
    A) 72   B) 63   C) 56   D) 54

    7. C

8. The sum of the first 4 digits of A = 12 = the sum of the last 3. The other (first 4, last 3) sums are B(17,18), C(19,18), D(16,15).
    A) 4 503 219   B) 4 526 819   C) 4 375 891   D) 6 073 915

    8. A

9. Perimeter $= 4 \times s$, so perimeter:side $= (4 \times s):(1 \times s) = 4:1$.
    A) 1:4   B) 1:2   C) 2:1   D) 4:1

    9. D

10. 1 quarter = 5 nickels, so (1999×1) quarters $= (1999 \times 5)$ nickels.
    A) $1999 \div 5$ B) 1999×5 C) $1999 + 5$ D) $1999 - 5$

    10. B

11. The numbers less than 20 from 79 are 60, 61, . . . , 79, . . . 97, 98.
    A) 19   B) 20   C) 38   D) 39

    11. D

12. A $\triangle$ has 3, a rect. & a sq. have 4, and a pentagon has 5 vertices.
    A) pentagon   B) square   C) rectangle   D) triangle

    12. A

13. I exercised on days 1, 3, 5, . . . , 363, 365. I exercised for 183 days. I didn't for 182.
    A) 182   B) 183   C) 365   D) 366

    13. B

14. In 12 345, the "1" means $1 \times 10\,000$, or 10^4.
    A) 10^2 B) 10^3 C) 10^4 D) 10^5

    14. C

15. $(10+20+30+40)$ cm $= 100$ cm $= 1$ m.
    A) 1 m   B) 10 m   C) 100 m   D) 1000 m

    15. A

16. One more day in March means $24 \times 60 = 1440$ more minutes.
    A) 720   B) 1440   C) 2160   D) 2880

    16. B

Go on to the next page ⫸ **6**

17. A 2×6 rectangle has perimeter 16. A 4×4 square has area 16.

A) 4   B) 12   C) 16   D) 144

17. C

18. Consider $4+5 = 9$. Only choice B (odd) works.

A) even   B) odd   C) prime   D) at least 10

18. B

19. $(19+\textbf{98}) \times \square = 19 \times \square + \textbf{98} \times \square$.

A) 19   B) 97   C) 98   D) 99

19. C

20. Mom bought a diving suit for $50 and sold it for $100. She made a profit of $50, 100% of her cost.

A) 50   B) 100   C) 150   D) 200

20. B

21. 1 isn't prime. $2+3+5+7+11+13+17+19+23+29 = 129$.

A) 101   B) 109   C) 110   D) 129

21. D

22. If the sum of the digits is 200, the number must have more than $200/9 \approx 22$ digits.

A) 10 000   B) 1000   C) 100   D) 10

22. D

23. A rope 2 m long can form a circle 2 m around.

A) 2 m   B) π m   C) 2π m   D) 4π m

23. A

24. The $\triangle$ has base 12 and height 8, so its area is $(1/2) \times 12 \times 8 = 48$. Rectangle $- \triangle = 96 - 48 = 48$.

A) 24   B) 48   C) 64   D) 96

24. B

25. If they were both born on Jan. 1 at noon, a woman in her 90's is *at most* $99 - 86 = 13$ years older than an 86-year-old.

A) 4   B) 10   C) 13   D) 14

25. C

26. The length of a side of a square is an even number. The perimeter of this square must be divisible by 8, so it could be 72.

A) 12   B) 30   C) 54   D) 72

26. D

27. When I hit it in 1999, it was Tues. When I hit it on Mar 9, 2000, it's 366 days later, which is 52 weeks and 2 days later; hence, it's a Thurs.

A) Monday   B) Tuesday
C) Wednesday   D) Thursday

27. D

28. Choice A $= 1999^2 + 1999 = 1999 \times (1999+1)$

A) $1999^2 + 1998 + 1$ B) $1999^2 + 1$
C) $1999^2 + 1998$ D) $1999^2 - 1$

28. A

29. A circle with $r = 2$ has perimeter 4π, as does a square of side π.

A) 1   B) π C) 4   D) 4π

29. B

Go on to the next page ⟫⟫ **6**

127

30.	For 2: 2+1, for 3: 3+1; for 5: 5+1. All primes! Only for primes is the sum of the whole number factors 1 more than the number. A) odd B) even C) prime D) composite	30. C
31.	Rewrite as $(2 \times 3) \times (2 \times 3) \times (2 \times 3) \times (64) = 6^3 \times 8^2$. A) $2^8 \times 3^3$ B) $4^2 \times 6^3$ C) $6^3 \times 8^2$ D) $2^4 \times 12^3$	31. C
32.	How many numbers between 1 and 500 are multiples of 2 but *not* 5? There are 249 even numbers, but remove the 49 multiples of 10. A) 50 B) 200 C) 249 D) 250	32. B
33.	The Town of *Us*, with 3400 people, gains 70 monthly. The town of *Them*, with 10 600 people, loses 130 monthly. Monthly, the populations are 200 closer. Population equality will take $(10\,600 - 3400) \div 200 = 36$ months. A) 36 B) 70 C) 72 D) 120	33. A
34.	Since $24 = 4 \times 3 \times 2 \times 1$, the answer is choice D. A) $27 \times 26 \times 25 \times \ldots \times 10 \times 9 \times 8$ B) $26 \times 25 \times 24 \times \ldots \times 9 \times 8 \times 7$ C) $25 \times 24 \times 23 \times \ldots \times 8 \times 7 \times 6$ D) $24 \times 23 \times 22 \times \ldots \times 7 \times 6 \times 5$	34. D
35.	My class was lined up on the gym floor in 9 rows, with 4 students in each row. If our coach rearranged us so that the number of rows was the same as the number of students in each row, we 36 students could form one 6×6 square. A) 8 B) 6 C) 5 D) 4	35. B
36.	$117 = 3 \times 3 \times 13$, $51 = 3 \times 17$. lcm $= (3 \times 3 \times 13) \times 17$. A) 1 B) 117×3 C) 117×17 D) 117×51	36. C
37.	When Pat wrote 100 numbers, each less than 100, on a notepad, Pat's magic pencil subtracted each number from 100, then wrote all 100 results on the notepad. Each pair (Pat's #, Pencil's #) has sum 100, so sum of 100 *pairs* $= 100 \times 100 = 10\,000$. A) 10 000 B) 20 000 C) 30 000 D) 40 000	37. A
38.	160 2's, so 32 groups of $2 \times 2 \times 2 \times 2 \times 2$, or 32 to the 32nd. A) 5th B) 10th C) 25th D) 32nd	38. D
39.	If a triangle's angles are perfect squares, they are 16°, 64°, and 100°. A) 4° B) 9° C) 16° D) 36°	39. C
40.	Each # after the first two is the sum of all previous numbers. Start with any two numbers. The pattern is: from 4th # on, the terms start doubling. Since 10th # = 1000, 9th is 500, and 8th is 250. A) 250 B) 500 C) 800 D) 1000	40. A

The end of the contest ✍ **6**

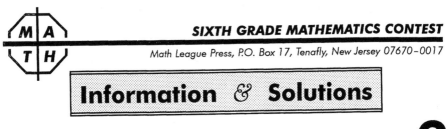

Information & Solutions

Tuesday, March 14, 2000

Contest Information

6

- **Solutions** Turn the page for detailed contest solutions (written in the question boxes) and letter answers (written in the *Answers* column to the right of each question).

- **Scores** Please remember that *this is a contest, not a test*—and there is no "passing" or "failing" score. Few students score as high as 30 points (75% correct). Students with half that, 15 points, *deserve commendation!*

- **Answers & Rating Scale** Turn to page 151 for the letter answers to each question and the rating scale for this contest.

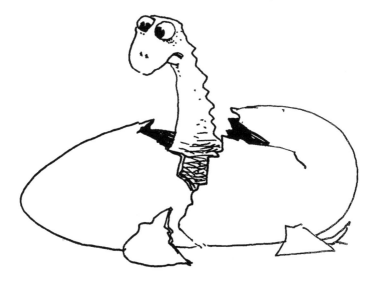

1. A = (2+34)+(1+56); B = (1+2+34)+56; and C = (1+34)+(2+56). A) 36 + 57 B) 37 + 56 C) 35 + 58 D) 37 + 59	1. D
2. Since 7 is a factor of 4×5×6×7×8×9×10×11×13, remainder is 0. A) 0 B) 3 C) 4 D) 6	2. A
3. Of 27 mice, 24 are not hurrying. Thus, 27 − 24 = 3 mice are hurrying. A) 2 B) 3 C) 12 D) 24	3. B
4. 11×(1 + 10 + 100) = 11×111 = 1221. A) 111 111 B) 12 321 C) 1221 D) 144	4. C
5. 2×4×6×8×10 has 5 more factors of 2 than 1×2×3×4×5. A) 2 B) 5 C) 16 D) 32 = 2^5	5. D
6. 2×(36+64) = (2×36) + (2×64) = 72 + 128. A) 2×36 + 64 B) 2 + 36 + 64 C) 36 + 64×2 D) 72 + 128	6. D
7. Every whole number is divisible by 1, but not by 0. A) 1, but not 0 B) 0, but not 1 C) 0 and 1 D) 0, 1, and 2	7. A
8. My bird sings 7×3 = 21 hrs each week, or 6×21 hrs in 6 weeks. A) 9 hours B) 18 hours C) 42 hours D) 126 hours	8. D
9. Choice D rounds to 60, since it is closer to 60 than to 50. A) 45.45 B) 50.50 C) 52.52 D) 55.55	9. D
10. The smallest difference comes from the smallest primes: 3−2 = 1. A) 1 B) 2 C) 3 D) 4	10. A
11. The same $9.60 is raised by 8 kids paying $1.20 or 6 kids paying $1.60, so the increase in each share is $1.60−$1.20 = 40¢. A) 15¢ B) 20¢ C) 30¢ D) 40¢	11. D
12. 230 million − 20 million = (230−20) million. A) 3 B) 21 C) 210 D) 228	12. C
13. 1 dino in 12 secs = 5 dinos in 1 min = 30×5 = 150 dinos every half-hour. A) 150 B) 216 C) 240 D) 360	13. A
14. 0 is even, since 0÷2 is a whole #. A) even B) odd C) prime D) negative	14. A
15. The 3-angle sum is 180°, so the average is 180°/3 = 60°. A) 45° B) 60° C) 90° D) 180°	15. B
16. Simplifies to "2 more than half itself." By trial and error, 4 works! A) 1 B) 2 C) 4 D) 8	16. C

Go on to the next page ⫸ **6**

130

17. The largest even factor of $6 \times 7 \times 8 \times 9 \times 10 = 30\,240$ is $30\,240$. A) 32 B) 160 C) 480 D) 30 240	17. D
18. Since the ratio of teeters to totters is $1:10 = 10:100 = 20:200$, it teeters 20 times for every 200 totters. A) 2000 B) 210 C) 20 D) 10	18. C
19. I watched TV for 37 hours = 1 day + 13 hrs. I stopped 13 hrs after noon, at 1 A.M. A) 11 A.M. B) 1 A.M. C) 12 P.M. D) 1 P.M.	19. B
20. 21 of 30 students didn't get A's, and $21:30 = 7:10 = 70\%$. A) 21% B) 65% C) 70% D) 79%	20. C
21. Each kid plays 4 games a day. After 1 day, each had played 4 games; after 2 days, 8 games; after 3 days, 12 games; after 4 days, 16 games; after 5 days, 20 games. # of games is a multiple of 4. A) 8 B) 12 C) 18 D) 20	21. C
22. Their average, $(0.25+4)/2 = 2.125$, is midway between them. A) 1.875 B) 2.0625 C) 2.125 D) 2.25	22. C
23. Cathy sells cakes for $9 each. It costs her $10 to make 4 cakes, so each cake costs her $10÷4 = $2.50, a per-cake profit of $9−$2.50. A) $0.25 B) $1.00 C) $2.50 D) $6.50	23. D
24. When I double 2, I get $2 \times 2 = 4$. When I square 2, I get $2^2 = 4$. A) 1 B) 2 C) 4 D) 16	24. B
25. I could dance on Sunday, Tuesday, Thursday, and Saturday. A) 2 B) 3 C) 4 D) 5	25. C
26. If Ann is 150 cm, then Bob is 152 cm, Carl is 153 cm, and Dee is tallest at 154 cm. Relative heights don't depend on Ann's true height. A) Ann B) Bob C) Carl D) Dee	26. D
27. If s is the length of one side of the square, then $s \times s \times s \times s = 256$, and $s = 4$ works. The sum $= 4+4+4+4 = 16$. A) 16 B) 32 C) 64 D) 256	27. A
28. 6 quarters = 150¢ & 9 dimes = 90¢; we need 60¢ = 12 nickels. A) 3 B) 12 C) 15 D) 30	28. B
29. The whole numbers are 0, 1, 2, 3, ... Choice A is the answer. A) 0 B) $1 = 0+1$ C) $2 = 0+2$ D) $100 = 0+100$	29. A

Go on to the next page ⟱ **6**

30. Since $6^3 = 2^3 \times 3^3$, and $4^5 = 2^{10} = 2^3 \times 2^7$, their g.c.f. is $2^3 = 8$. A) 8 B) 12 C) 16 D) 136	30. A
31. If I take 3 socks, I might take 1 white, 1 black, and 1 blue sock. When I do that, the next sock I take must match one of these 3. A) 3 B) 4 C) 13 D) 14	31. B
32. Avg cost per pen = (tot cost) ÷ (tot # of pens). Tot cost = \$12 + \$12 = 2400¢. Tot # of pens = 1200¢ ÷ 40¢ + 1200¢ ÷ 60¢ = 30 + 20 = 50; avg cost = 2400¢ ÷ 50 = 48¢. A) 45¢ B) 48¢ C) 50¢ D) 52¢	32. B
33. Factoring, $2^3 \times 3^3 \times 4^3 \times (6^3) \times 9^3 =$ $2^3 \times 3^3 \times 2^6 \times (2^3 \times 3^3) \times 3^6 = 2^{12} \times 3^{12}$. A) $2^9 \times 3^9$ B) $2^{12} \times 3^{12}$ C) $2^{15} \times 3^{15}$ D) $2^{54} \times 3^{54}$	33. B
34. As in #26, try an example. If my time is 100 secs, and Pat's is 120 secs, then your time is 108 secs. You beat Pat by 12 seconds. A) 6 seconds B) 12 seconds C) 18 seconds D) 24 seconds	34. B
35. Possible ratios: **1:11**; 2:10 = **1:5**; 3:9 = **1:3**; 4:8 = **1:2**; **5:7**; 6:6 = **1:1**. A) 1:2 B) 1:3 C) 1:4 D) 1:5	35. C
36. Find difference between each even # and the odd # just below: $(2-1) + (4-3) + \ldots + (20\,000 - 19\,999) = 1 + 1 + \ldots + 1 = 10\,000$. A) 1 B) 5000 C) 9999 D) 10 000	36. D
37. Area of square = $4 \times$ shaded part = $4 \times 16 = 64$; side of square = $\sqrt{64} = 8$; perimeter = $4 \times 8 = 32$. A) 8 B) 16 C) 32 D) 64	37. C
38. 2000 ÷ 1001 and 2000 ÷ 999 leave whole # remainders 999 and 2. A) even B) odd C) prime D) whole	38. D
39. Factor: $2^{1998} \times (2^3 - 2^2 - 2 - 1) = 2^{1998}$. A) 2 B) 2^4 C) 2^{500} D) 2^{1998}	39. D
40. Clark ran as far in 15 seconds as Lois ran in 10 seconds. Since Lois ran 60 m in 10 seconds, Clark ran 60 m in 15 seconds. Clark's average running speed was 60 m ÷ 15 seconds = 4 m/sec. A) 2 m/sec B) 4 m/sec C) 6 m/sec D) 14 m/sec	40. B

The end of the contest ✍ **6**

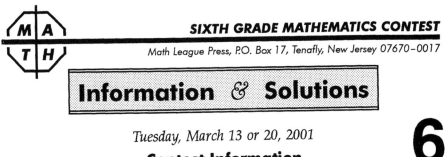
Information & Solutions

Tuesday, March 13 or 20, 2001

Contest Information

6

- **Solutions** Turn the page for detailed contest solutions (written in the question boxes) and letter answers (written in the *Answers* column to the right of each question).

- **Scores** Please remember that *this is a contest, not a test*—and there is no "passing" or "failing" score. Few students score as high as 30 points (75% correct). Students with half that, 15 points, *deserve commendation!*

- **Answers & Rating Scale** Turn to page 152 for the letter answers to each question and the rating scale for this contest.

1. $77 - (33 + 44) = 0 = 77 - 33 - 44 = (77 - 33) - 44$. A) 44　　　B) 33　　　C) 22　　　D) 0	1. A
2. Kat the lion tamer tames 9 lions every year. In 9 years, Kat tames $9 \times 9 = 81$ lions. A) 18　　B) 81　　C) 90　　D) 99	2. B
3. To find original #, work backwards, reversing both operations: $24 - 6 = 18$, and $18 \times 2 = 36$. A) 9　　B) 15　　C) 36　　D) 54	3. C
4. The only division with a remainder is choice B. A) $72 \div 36 = 2$　B) $218 \div 18$　　C) $416 \div 16 = 26$ D) $616 \div 14 = 44$	4. B
5. The # that's 1 more than 1 million is $1\,000\,001$. The # that's 1 less than 1 million is $999\,999$. Subtract to get 2. A) 1　　　B) 2　　　C) $1\,000\,000$　　D) $1\,000\,001$	5. B
6. Convert to years: $120 \div 12 = 10$ years. Their age sum is $5 \times 10 = 50$ years. A) 10　　B) 24　　C) 50　　D) 120	6. C
7. Add the digits to test: for 603, $6 + 0 + 3 = 9$. A) 663　　B) 603　　C) 336　　D) 303	7. B
8. The difference is either $(366 - 29)$ or $(365 - 28)$. Each equals 337. A) 335　　　B) 336　　　C) 337　　　D) 338	8. C
9. A typical gorilla eats 2 dozen bananas daily, and $2 \times 12 = 24$. Ten typical gorillas eat $10 \times 24 = 240$ bananas daily. A) 12　　　B) 20　　　C) 200　　　D) 240	9. D
10. The tens' digit of 11×11 is a 2. The other digits have no effect. A) 0　　　B) 1　　　C) 2　　　D) 3	10. C
11. Every prime has itself and 1 as factors. Note that 1 is not a prime. A) 0　　　B) 1　　　C) 2　　　D) 3	11. C
12. The helicopter can fly for 1½ hours on 1 full tank of gas. Double, then double again to see that it can fly for 6 hours on 4 full tanks. A) 3　　　B) 4　　　C) 15　　　D) 60	12. B
13. From 1st to 51st is 51 people, and 49 people stand *between* the two ends. A) 49　B) 50　C) 51　D) 52	13. A
14. Given $= 2^{2+3+4+5} = 2^{14} = (2^2)^7$. A) 60　B) 2^6　C) 240　D) 4^7	14. D
15. Eliminate choices. Try *several* examples. A good example is $7 \div 4$. A) odd　　　B) even　　　C) 1　　　D) prime	15. A

Go on to the next page ▶ **6**

16. The greatest common factor of $8^2 = 64$ and $2^8 = 256$ is $64 = 4^3$. A) 2^2 B) 2^3 C) 4^2 D) 4^3	16. D
17. To round 249 973 to the thousands' place, look at the 3rd digit from the right, a 9. Round *UP* to 250 000. Only one digit is unchanged. A) 1 B) 2 C) 3 D) 5	17. A
18. Since $300\text{g} \div 6\text{ kg} = 300\text{ g} \div 6000\text{ g} = 0.05$, the answer is 5%. A) 2 B) 5 C) 50 D) 5000	18. B
19. Besides Dave, there are 7 other kids. If 2 kids have dived since Dave dove, he must wait for 5 more to dive before he dives again. A) 3 B) 4 C) 5 D) 6	19. C
20. $\sqrt{100-64} + \sqrt{25-16} = 6+3 = 9 = \sqrt{81}$. A) $\sqrt{81}$ B) $\sqrt{45}$ C) $\sqrt{9}$ D) -65	20. A
21. The largest *prime* factors of 6, 18, & 30 are 3, 3, & 5 respectively. A) 2 B) 3 C) 5 D) 6^3	21. C
22. The number whose square is 16 is 4. The square root of 4 is 2. A) 2 B) 4 C) $\sqrt{8}$ D) 16	22. A
23. Leap year = 366 days; 52 weeks = $52 \times 7 = 364$ days. Subtract. A) 1 day B) 2 days C) 3 days D) 1 week	23. B
24. One angle is 40°, so the other two total 140°. Their average is 70°. A) 70° B) 140° C) 160° D) 180°	24. A
25. Convert dimensions: 12 km/hr = 12 000 m/hr = 12 000 m/(60 minutes) = 200 m/minute. A) 20 B) 72 C) 200 D) 720	25. C
26. "2," an even whole prime, is divisible by 2. A) even B) prime C) whole D) odd	26. D
27. 8 rotations = 8 m, so 1 rotation = 1 m = $2\pi r$, and $r = 1$ m $\div (2\pi)$. A) 1 B) $\pi \div 2$ C) 2 D) $1 \div (2\pi)$	27. D
28. In 10 hours, my watch gained $(3 + 6 + 9 + 12 + 15 + 18 + 21 + 24 + 27 + 30)$ minutes = 165 minutes. A) 30 minutes B) 135 minutes C) 165 minutes D) 198 minutes	28. C
29. $(\text{Side})^4 = 1296$; $(\text{side})^2 = \sqrt{1296} = 36$; side = $\sqrt{36} = 6$. Sum = 4×6. A) 24 B) 36 C) 48 D) 72	29. A
30. The only positive whole number whose square is equal to its square root is 1, so the answer is B. A) 0 B) 1 C) 2 D) 4	30. B

Go on to the next page ▐▌▶ **6**

135

31. Since $2 \times 60 = 120$, Ali is older than 120 months. Therefore, only choice D is possible. In fact, Ali is 12 years old today.
    A) 12 months   B) 24 months   C) 120 months   D) 144 months

    31.
    D

32. Factoring into primes, $1001 = 7 \times 11 \times 13$. Consequently, Britney spears $(7 + 11 + 13)$ shrimp = 31 shrimp.

    A) 31   B) 37   C) 103   D) 151

    32.
    A

33. Rabbit *could have* run 1.5 km in 6.0 mins., so he ran 1 km in 4.0 mins.
    A) 3.0 minutes   B) 4.0 minutes
    C) 4.5 minutes   D) 9.0 minutes

    33.
    B

34. The 5 consecutive integers have the sum $11+12+\ldots+20 = 155$, so $155 \div 5 = 31$ is the middle integer of $29+30+31+32+33 = 155$.
    A) 22   B) 29   C) 31   D) 33

    34.
    B

35. Here, dividend = (quotient) × divisor = (divisor) × divisor.
    A) $\sqrt{\text{divisor}}$ B) divisor   C) divisor^2 D) quotient

    35.
    C

36. Reading each backwards, we would get 9886, 1991, 1691, and 1881. Upside down, 1991 becomes 1661, not 1991.
    A) 6889   B) 1991   C) 1961   D) 1881

    36.
    D

37. Largest difference = $(50+49+\ldots+2+1) - (44+43+\ldots+1+0) = 50 + 49 + 48 + 47 + 46 + 45 = 285$.
    A) 10   B) 15   C) 240   D) 285

    37.
    D

38. Find a pattern: 1st # = 1, 2nd = $1 \times 3 = 3^1$, 3rd = $3 \times 3^1 = 3^2$, 4th = $3 \times 3^2 = 3^3$, 5th = $3 \times 3^3 = 3^4, \ldots$, 200th number = 3^{199}.
    A) 600   B) 900   C) 3^{199} D) 3^{200}

    38.
    C

39. Rewrite ratios so # of dimes is same in both. Ratio of nickels to dimes is 2:3 = 16:24. Ratio of dimes to dollars is 8:1 = 24:3. For every 16 nickels, there are 24 dimes (for $24+16 = 40$ coins) and 3 dollars.
    A) 40:3   B) 16:3   C) 13:1   D) 5:1

    39.
    A

40. $20 \times (19 \times 18 \times \ldots \times 2 \times 1) - 1 \times (19 \times 18 \times \ldots \times 2 \times 1) = (20 - 1) \times (19 \times 18 \times \ldots \times 2 \times 1] = 19 \times 19 \times 18 \times 17 \times \ldots \times 3 \times 2 \times 1$
    A) 19   B) $19 \times 18 \times 17 \times \ldots \times 3 \times 2 \times 1$
    C) 20   D) $19 \times 19 \times 18 \times 17 \times \ldots \times 3 \times 2 \times 1$

    40.
    D

The end of the contest ✍ **6**

136

Answer Keys & Difficulty Ratings

1996-1997 through 2000-2001

ANSWERS, 1996-97 4th Grade Contest

1. B	7. A	13. C	19. A	25. D
2. D	8. A	14. C	20. A	26. B
3. C	9. B	15. D	21. B	27. A
4. B	10. C	16. A	22. B	28. C
5. D	11. A	17. C	23. B	29. B
6. D	12. C	18. D	24. C	30. C

RATE YOURSELF!!!
for the 1996-97 4th GRADE CONTEST

Score	Rating
28-30	Another Einstein
25-27	Mathematical Wizard
22-24	School Champion
18-21	Grade Level Champion
16-17	Best In The Class
14-15	Excellent Student
11-13	Good Student
9-10	Average Student
0-8	Better Luck Next Time

ANSWERS, 1997-98 4th Grade Contest

1. B	7. A	13. D	19. B	25. A
2. D	8. D	14. B	20. D	26. C
3. B	9. C	15. B	21. B	27. A
4. A	10. D	16. D	22. C	28. C
5. C	11. A	17. C	23. A	29. D
6. C	12. B	18. C	24. B	30. C

RATE YOURSELF!!!
for the 1997-98 4th GRADE CONTEST

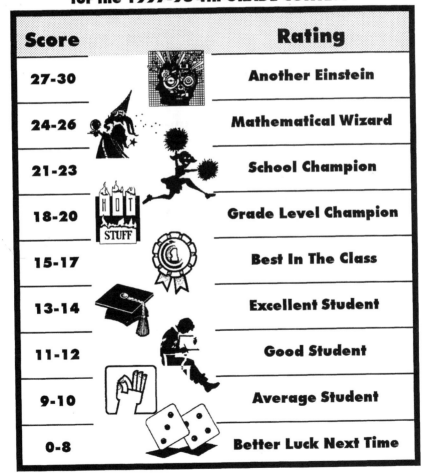

Score	Rating
27-30	Another Einstein
24-26	Mathematical Wizard
21-23	School Champion
18-20	Grade Level Champion
15-17	Best In The Class
13-14	Excellent Student
11-12	Good Student
9-10	Average Student
0-8	Better Luck Next Time

ANSWERS, 1998-99 4th Grade Contest

1. B	7. C	13. B	19. A	25. D
2. D	8. C	14. A	20. B	26. A
3. B	9. C	15. C	21. D	27. B
4. D	10. B	16. C	22. D	28. A
5. B	11. A	17. A	23. A	29. D
6. A	12. C	18. B	24. C	30. B

RATE YOURSELF!!!
for the 1998-99 4th GRADE CONTEST

Score	Rating
28-30	Another Einstein
26-27	Mathematical Wizard
23-25	School Champion
21-22	Grade Level Champion
19-20	Best In The Class
16-18	Excellent Student
13-15	Good Student
10-12	Average Student
0-9	Better Luck Next Time

ANSWERS, 1999-00 4th Grade Contest

1. C	7. A	13. A	19. C	25. A
2. C	8. C	14. C	20. C	26. D
3. D	9. A	15. B	21. B	27. C
4. D	10. D	16. A	22. B	28. B
5. A	11. C	17. C	23. B	29. C
6. A	12. B	18. D	24. C	30. B

RATE YOURSELF!!!
for the 1999-00 4th GRADE CONTEST

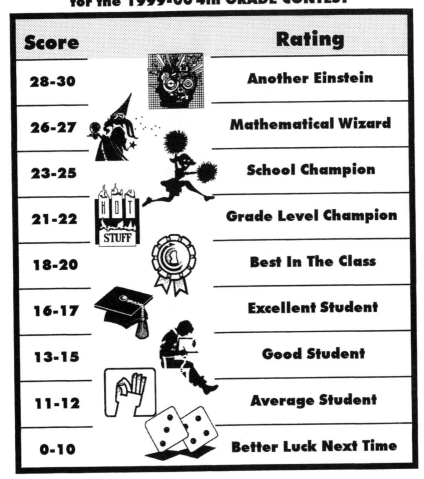

Score	Rating
28-30	Another Einstein
26-27	Mathematical Wizard
23-25	School Champion
21-22	Grade Level Champion
18-20	Best In The Class
16-17	Excellent Student
13-15	Good Student
11-12	Average Student
0-10	Better Luck Next Time

ANSWERS, 2000-01 4th Grade Contest

1. C	7. B	13. B	19. B	25. D
2. D	8. B	14. C	20. C	26. C
3. B	9. A	15. D	21. A	27. B
4. D	10. A	16. C	22. D	28. D
5. C	11. A	17. A	23. A	29. B
6. C	12. C	18. D	24. D	30. B

RATE YOURSELF!!!
for the 2000-01 4th GRADE CONTEST

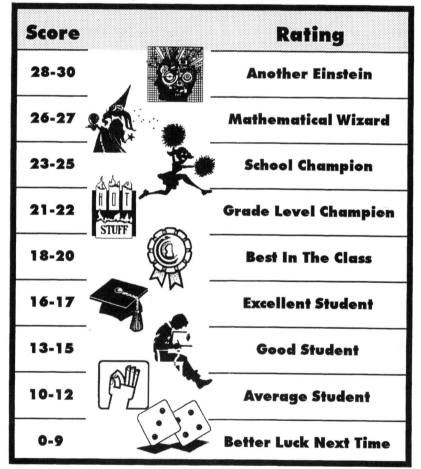

Score	Rating
28-30	Another Einstein
26-27	Mathematical Wizard
23-25	School Champion
21-22	Grade Level Champion
18-20	Best In The Class
16-17	Excellent Student
13-15	Good Student
10-12	Average Student
0-9	Better Luck Next Time

ANSWERS, 1996-97 5th Grade Contest

1. B	7. C	13. A	19. B	25. D
2. A	8. D	14. A	20. D	26. A
3. B	9. B	15. D	21. B	27. C
4. A	10. C	16. B	22. C	28. B
5. A	11. D	17. D	23. A	29. C
6. C	12. C	18. A	24. D	30. B

RATE YOURSELF!!!
for the 1996-97 5th GRADE CONTEST

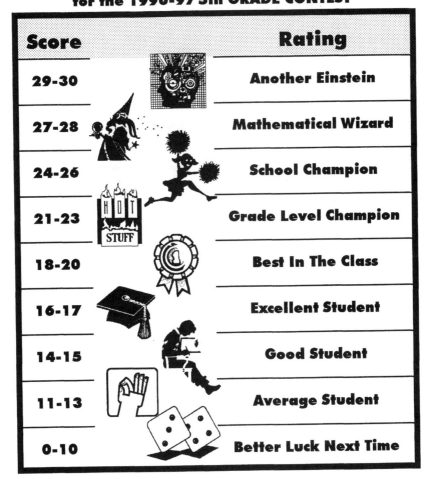

Score	Rating
29-30	Another Einstein
27-28	Mathematical Wizard
24-26	School Champion
21-23	Grade Level Champion
18-20	Best In The Class
16-17	Excellent Student
14-15	Good Student
11-13	Average Student
0-10	Better Luck Next Time

ANSWERS, 1997-98 5th Grade Contest

1. D	7. A	13. C	19. A	25. A
2. D	8. B	14. C	20. B	26. C
3. C	9. D	15. B	21. D	27. C
4. A	10. C	16. A	22. D	28. B
5. B	11. D	17. B	23. B	29. A
6. C	12. A	18. C	24. D	30. A

RATE YOURSELF!!!
for the 1997-98 5th GRADE CONTEST

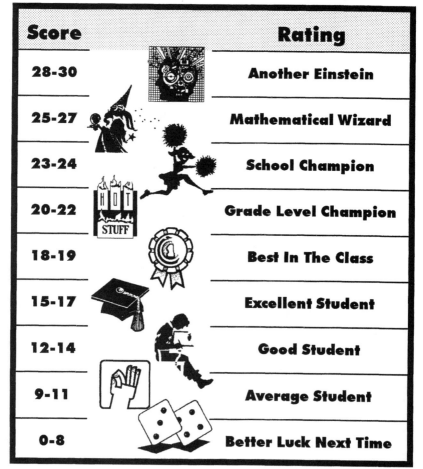

Score	Rating
28-30	Another Einstein
25-27	Mathematical Wizard
23-24	School Champion
20-22	Grade Level Champion
18-19	Best In The Class
15-17	Excellent Student
12-14	Good Student
9-11	Average Student
0-8	Better Luck Next Time

ANSWERS, 1998-99 5th Grade Contest

1. D	7. C	13. B	19. B	25. D
2. B	8. B	14. A	20. D	26. B
3. C	9. A	15. D	21. A	27. B
4. A	10. C	16. D	22. B	28. A
5. D	11. D	17. A	23. C	29. A
6. B	12. A	18. C	24. C	30. D

RATE YOURSELF!!!
for the 1998-99 5th GRADE CONTEST

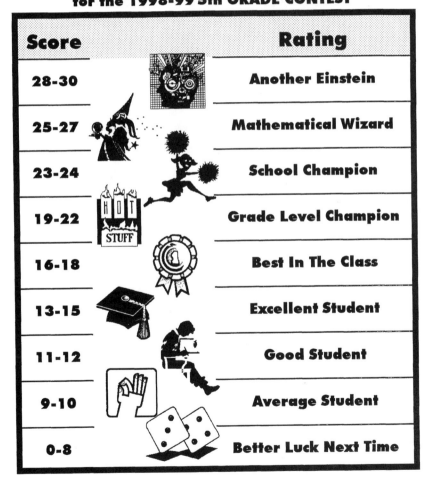

Score	Rating
28-30	Another Einstein
25-27	Mathematical Wizard
23-24	School Champion
19-22	Grade Level Champion
16-18	Best In The Class
13-15	Excellent Student
11-12	Good Student
9-10	Average Student
0-8	Better Luck Next Time

ANSWERS, 1999-00 5th Grade Contest

1. D	7. C	13. D	19. B	25. A
2. C	8. A	14. B	20. B	26. D
3. B	9. A	15. D	21. C	27. D
4. A	10. A	16. A	22. C	28. C
5. D	11. D	17. B	23. B	29. B
6. D	12. D	18. D	24. D	30. A

RATE YOURSELF!!!
for the 1999-00 5th GRADE CONTEST

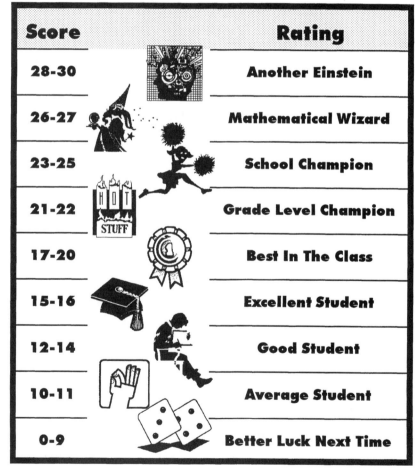

Score	Rating
28-30	Another Einstein
26-27	Mathematical Wizard
23-25	School Champion
21-22	Grade Level Champion
17-20	Best In The Class
15-16	Excellent Student
12-14	Good Student
10-11	Average Student
0-9	Better Luck Next Time

ANSWERS, 2000-01 5th Grade Contest

1. B	7. A	13. B	19. A	25. B
2. D	8. C	14. C	20. A	26. B
3. D	9. D	15. C	21. B	27. B
4. A	10. A	16. B	22. D	28. A
5. D	11. C	17. C	23. C	29. D
6. B	12. C	18. B	24. B	30. C

RATE YOURSELF!!!
for the 2000-01 5th GRADE CONTEST

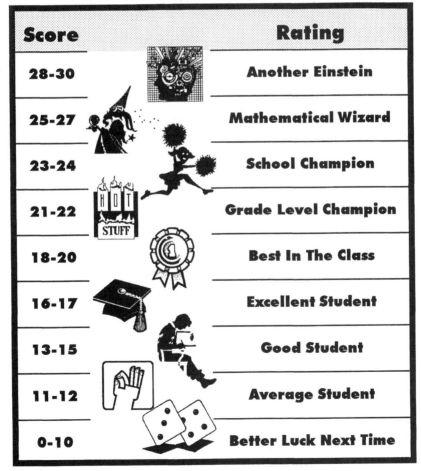

Score	Rating
28-30	Another Einstein
25-27	Mathematical Wizard
23-24	School Champion
21-22	Grade Level Champion
18-20	Best In The Class
16-17	Excellent Student
13-15	Good Student
11-12	Average Student
0-10	Better Luck Next Time

ANSWERS, 1996-97 6th Grade Contest

1. D	9. A	17. D	25. C	33. C
2. C	10. A	18. A	26. D	34. B
3. D	11. A	19. B	27. A	35. D
4. C	12. B	20. B	28. C	36. C
5. B	13. C	21. D	29. A	37. A
6. B	14. C	22. C	30. D	38. B
7. D	15. B	23. D	31. D	39. C
8. B	16. B	24. C	32. A	40. B

RATE YOURSELF!!!
for the 1996-97 6th GRADE CONTEST

Score	Rating
38-40	Another Einstein
35-37	Mathematical Wizard
32-34	School Champion
28-31	Grade Level Champion
25-27	Best In The Class
21-24	Excellent Student
18-20	Good Student
15-17	Average Student
0-14	Better Luck Next Time

ANSWERS, 1997-98 6th Grade Contest

1. C	9. B	17. D	25. A	33. B
2. C	10. B	18. A	26. B	34. C
3. D	11. C	19. D	27. D	35. A
4. C	12. D	20. C	28. B	36. C
5. B	13. A	21. B	29. B	37. B
6. A	14. A	22. D	30. C	38. A
7. B	15. D	23. D	31. D	39. C
8. B	16. A	24. C	32. A	40. A

RATE YOURSELF!!!
for the 1997-98 6th GRADE CONTEST

Score	Rating
38-40	Another Einstein
35-37	Mathematical Wizard
31-34	School Champion
28-30	Grade Level Champion
24-27	Best In The Class
21-23	Excellent Student
17-20	Good Student
13-16	Average Student
0-12	Better Luck Next Time

ANSWERS, 1998-99 6th Grade Contest

1. D	9. D	17. C	25. C	33. A
2. D	10. B	18. B	26. D	34. D
3. A	11. D	19. C	27. D	35. B
4. D	12. A	20. B	28. A	36. C
5. C	13. B	21. D	29. B	37. A
6. B	14. C	22. D	30. C	38. D
7. C	15. A	23. A	31. C	39. C
8. A	16. B	24. B	32. B	40. A

RATE YOURSELF!!!
for the 1998-99 6th GRADE CONTEST

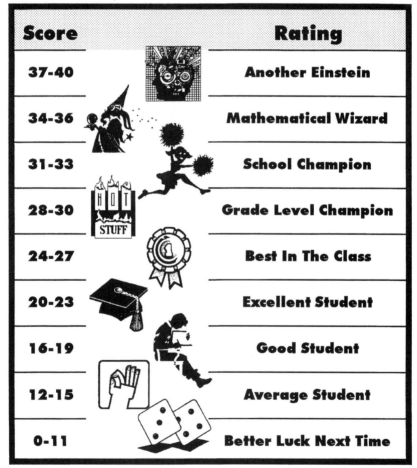

Score	Rating
37-40	Another Einstein
34-36	Mathematical Wizard
31-33	School Champion
28-30	Grade Level Champion
24-27	Best In The Class
20-23	Excellent Student
16-19	Good Student
12-15	Average Student
0-11	Better Luck Next Time

ANSWERS, 1999-00 6th Grade Contest

1. D	9. D	17. D	25. C	33. B
2. A	10. A	18. C	26. D	34. B
3. B	11. D	19. B	27. A	35. C
4. C	12. C	20. C	28. B	36. D
5. D	13. A	21. C	29. A	37. C
6. D	14. A	22. C	30. A	38. D
7. A	15. B	23. D	31. B	39. D
8. D	16. C	24. B	32. B	40. B

RATE YOURSELF!!!
for the 1999-00 6th GRADE CONTEST

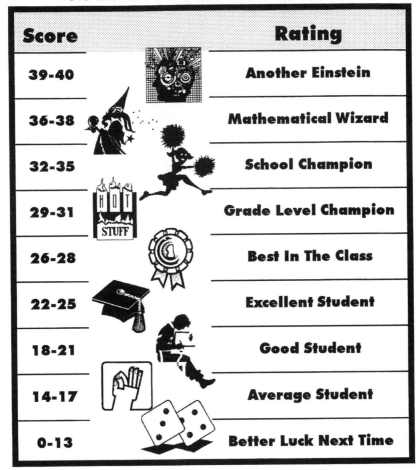

Score	Rating
39-40	Another Einstein
36-38	Mathematical Wizard
32-35	School Champion
29-31	Grade Level Champion
26-28	Best In The Class
22-25	Excellent Student
18-21	Good Student
14-17	Average Student
0-13	Better Luck Next Time

ANSWERS, 2000-01 6th Grade Contest

1. A	9. D	17. A	25. C	33. B
2. B	10. C	18. B	26. D	34. B
3. C	11. C	19. C	27. D	35. C
4. B	12. B	20. A	28. C	36. D
5. B	13. A	21. C	29. A	37. D
6. C	14. D	22. A	30. B	38. C
7. B	15. A	23. B	31. D	39. A
8. C	16. D	24. A	32. A	40. D

RATE YOURSELF!!!
for the 2000-01 6th GRADE CONTEST

Score	Rating
38-40	Another Einstein
35-37	Mathematical Wizard
32-34	School Champion
29-31	Grade Level Champion
25-28	Best In The Class
21-24	Excellent Student
17-20	Good Student
13-16	Average Student
0-12	Better Luck Next Time

Math League Contest Books
4th Grade Through High School Levels

Written by Steven R. Conrad and Daniel Flegler, recipients of President Reagan's 1985 Presidential Awards for Excellence in Mathematics Teaching, each book provides schools and students with:

- *Easy-to-use format designed for a 30-minute period*
- *Problems ranging from straightforward to challenging*

Use the form below (or a copy) to order your books

Name: _____

Address: _____

City: _____ State: _____ Zip: _____
        (or Province)   (or Postal Code)

Available Titles	**# of Copies**	**Cost**
Math Contests—Grades 4, 5, 6	($12.95 each, $15.95 Canadian)	
Volume 1: 1979-80 through 1985-86	_____	_____
Volume 2: 1986-87 through 1990-91	_____	_____
Volume 3: 1991-92 through 1995-96	_____	_____
Volume 4: 1996-97 through 2000-01	_____	_____
Volume 5: 2001-02 through 2005-06	_____	_____
Math Contests—Grades 7 & 8 ‡	‡(Vols. 3,4,5 include Alg. Course I)	
Volume 1: 1977-78 through 1981-82	_____	_____
Volume 2: 1982-83 through 1990-91	_____	_____
Volume 3: 1991-92 through 1995-96	_____	_____
Volume 4: 1996-97 through 2000-01	_____	_____
Volume 5: 2001-02 through 2005-06	_____	_____
Math Contests—High School		
Volume 1: 1977-78 through 1981-82	_____	_____
Volume 2: 1982-83 through 1990-91	_____	_____
Volume 3: 1991-92 through 1995-96	_____	_____
Volume 4: 1996-97 through 2000-01	_____	_____
Volume 5: 2001-02 through 2005-06	_____	_____
Shipping and Handling	**$3 ($5 Canadian)**	

Please allow 4-6 weeks for delivery   Total: $_____

☐ Check or Purchase Order Enclosed; **or**

☐ Visa / MasterCard/Discover # _____

☐ Exp. Date _____ Signature _____

Mail your order with payment to:
Math League Press. PO Box 17, Tenafly, New Jersey USA 07670–0017
or order on the Web at www.mathleague.com

Phone: (201) 568-6328 • Fax: (201) 816-0125